Updates in Surgery

The aim of this series is to provide informative updates on hot topics in the areas of breast, endocrine, and abdominal surgery, surgical oncology, and coloproctology, and on new surgical techniques such as robotic surgery, laparoscopy, and minimally invasive surgery. Readers will find detailed guidance on patient selection, performance of surgical procedures, and avoidance of complications. In addition, a range of other important aspects are covered, from the role of new imaging tools to the use of combined treatments and postoperative care.

The topics addressed by volumes in the series Updates in Surgery have been selected for their broad significance in collaboration with the Italian Society of Surgery. Each volume will assist surgical residents and fellows and practicing surgeons in reaching appropriate treatment decisions and achieving optimal outcomes. The series will also be highly relevant for surgical researchers.

Guido A. M. Tiberio
Editor

Primary Adrenal Malignancies

 Springer

Editor
Guido A. M. Tiberio
General Surgery, Department of Clinical and Experimental Sciences
University of Brescia at ASST Spedali Civili di Brescia
Brescia, Italy

ISSN 2280-9848   ISSN 2281-0854   (electronic)
Updates in Surgery

ISBN 978-3-031-62300-4   ISBN 978-3-031-62301-1   (eBook)
https://doi.org/10.1007/978-3-031-62301-1

The publication and the distribution of this volume have been supported by the Italian Society of Surgery.

The editor of the volume and the Italian Society of Surgery would like to thank Retex S.p.A. Benefit Company for its unconditional contribution, which made it possible to publish this book under the Open Access model.

This book is an open access publication.

Revision and editing: R. M. Martorelli, Scienzaperta (Novate Milanese, Italy)

This Springer imprint is published by the registered company Springer Nature Switzerland AG
The registered company address is: Gewerbestrasse 11, 6330 Cham, Switzerland

If disposing of this product, please recycle the paper.

Foreword

The spirit of the Italian Society of Surgery in assigning the task of writing its biennial reports has always been to entrust great experts to deal with surgical topics of great relevance. Such is the case with this excellent monograph edited by Guido Tiberio, which, after Favia's 2000 Italian-language report on adrenal surgery in general, now addresses primary adrenal malignancies for the first time for our society.

It is a privilege for me to present this remarkable volume, which is an extremely valuable analysis of all aspects of these diseases. In its 20 chapters, it discusses adrenocortical carcinomas, pheochromocytomas, paragangliomas, and the so-called incidentalomas: epidemiological aspects, diagnosis, endocrine and hereditary syndromes, staging, prognostic factors, histopathology, genetics, molecular biology, imaging, open and minimally invasive surgery, medical and integrated treatments, up to radionuclide therapy and translational research.

Written in a clear and informative style, the precise and well-structured chapters of this important monograph answer the many questions that these malignancies pose. The high scientific level of the monograph makes it valuable for both experienced and younger surgeons, and I am sure the volume will be a reference for all those who want to delve into this intriguing topic.

Also on behalf of the Italian Society of Surgery, I want to sincerely thank and congratulate Tiberio and the co-authors of this book on their excellent work. I would also like to thank Springer for its organizational efficiency and editorial expertise in assisting my distinguished colleague in the production of this scientific publication and for enabling its worldwide dissemination, both in printed and open-access forms.

Italian Society of Surgery   Massimo Carlini
Rome, Italy
September 2024

Preface

Had I not been working at Brescia University and its referral hospital, Spedali Civili di Brescia, I would not have had the opportunity to edit this book. By chance or by necessity, in fact, a team of dedicated clinicians, intrigued by adrenal pathologies, set up an active clinical and research collaboration, and this book bears witness to that collaboration. Primary adrenal malignancies are among the rarest of neoplasms; hidden away among the more common benign adrenal tumors, they may easily go unrecognized and therefore not be managed in the appropriate manner. For these reasons and because of their biological aggressiveness, they carry a poor prognosis, also when detected in the early stages. The best and only strategy capable of improving the clinical management of rare and ultra-rare malignancies is to share and spread knowledge, which is the aim of this publication.

In this book, we shed light, in a user-friendly way, on all the different issues concerning adrenocortical carcinoma and malignant pheochromocytoma. Methodologically, we employed a transversal approach typical of modern precision medicine. In this way, all relevant clinical issues and their interconnections could be made readily available to the reader.

In our discipline, the equation yielding the best results has multiple components: knowledge is the main component, and centralization is the second one, since the competencies required for achieving a comprehensive approach to these rare diseases are closely related to caseloads. In modern terms, centralization should be defined as collaborative centralization, implying that the referring clinicians should actively participate in their patient's management in the tertiary institution, which also includes being involved in the surgical activity.

If the reader finds some practical use for our work, our mission will be accomplished, and the generous, unconditional contribution of Mr. Fausto Caprini, CEO of Retex S.p.A. Benefit Company, that allowed the open-access publication of this book, will have been spent in the best possible way.

Finally, we cannot forget our patients and their suffering. They are the focus of our interest and their trust and patience, confirmed over the course of time, is a strong source of support to our efforts.

Brescia, Italy
September 2024

Guido A. M. Tiberio

Contents

Epidemiology, Presentation, Staging, and Prognostic Factors in Adrenocortical Carcinoma

1

Deborah Cosentini, Valentina Cremaschi,
Salvatore Grisanti, Alfredo Berruti, and Marta Laganà

1.1 Epidemiology

The estimated incidence of adrenocortical carcinoma (ACC) is between 0.5 and 2 new cases per million per year in Western countries. The male/female ratio is 1/1.5 and, according to age, there is a bimodal distribution with two peaks in childhood and between the fourth and fifth decade [1].

1.2 Clinical Presentation

The majority of ACCs, about 50–60%, are functioning at presentation. Cushing's syndrome (hypercortisolism) or mixed Cushing's and virilizing syndromes are observed in 50–80% of hormone-secreting ACCs. Instead, pure androgen excess is less frequent and estrogen or mineralocorticoid excesses are very rare [2, 3]. The simultaneous presence of multiple secretions is typical of malignant diseases. Possible syndromes, sign and symptoms are summarized in Table 1.1.

Many ACC are discovered incidentally. With the use of modern imaging techniques, the so-called adrenal incidentalomas are increasingly detected and, based on the published literature, the frequencies of the different underlying tumor types are adrenocortical adenomas in 80%, ACC in 8%, pheochromocytomas in 7%, and metastatic tumors in 5% [4]. The proportion of malignancy is about 2% if adrenal incidentalomas are less than 4 cm, whereas it increases to 6% if they are 4–6 cm in size and up to 25% among those >6 cm [3, 4].

D. Cosentini (✉) · V. Cremaschi · S. Grisanti · A. Berruti · M. Laganà
Medical Oncology, Department of Medical and Surgical Specialties, Radiological Sciences, and Public Health, University of Brescia at ASST Spedali Civili di Brescia, Brescia, Italy
e-mail: deborah.cosentini@unibs.it; v.cremaschi@unibs.it; salvatore.grisanti@unibs.it; alfredo.berruti@unibs.it; marta.lagana@unibs.it

© The Author(s) 2025

1

G. A. M. Tiberio (ed.), *Primary Adrenal Malignancies*, Updates in Surgery, https://doi.org/10.1007/978-3-031-62301-1_1

Table 1.1 Syndromes, signs and symptoms related to hormone-secreting adrenocortical carcinomas

Syndrome	Incidence	Hormone profile	Signs and symptoms
Cushing's syndrome	50–80% of cases	Hypercortisolism, suppressed ACTH levels	Plethora, dorsal fat hump, diabetes mellitus, muscle weakness/atrophy, osteoporosis, hypokalemia, hypertension, mood alterations, insomnia, skin atrophy, higher susceptibility to infectious diseases
Hyperandrogenism	40–60% of cases	Excess of: DHEAS, 17-OHP, testosterone, androstenedione	In women: hirsutism, virilization, menstrual irregularities, temporal balding, acne
Hyperestrogenism	1–3% of cases	Estrogen excess	In men: gynecomastia and testicular atrophy
Hyperaldosteronism	2% of cases	Aldosterone excess	Hypokalemia, hypertension

ACHT adrenocorticotropic hormone, *DHEAS* dehydroepiandrosterone, *17-OHP* hydroxyprogesterone

Moreover, mass symptoms, including abdominal discomfort, nausea, vomiting, abdominal fullness or back pain, are present in about 30–40% of ACC patients at diagnosis [3].

The initial evaluation of a patient with ACC should include physical examination, patient history collection and imaging assessment. In particular, in cases of suspected ACC, an extensive steroid hormone work-up is recommended, assessing gluco-, mineralo-, sex-, and precursor-steroids [3]. In addition, for all adrenal masses, plasma-free or urinary-fractionated metanephrines should be measured in order to exclude pheochromocytoma. The aims of hormonal evaluation are multiple: a differential diagnosis with orientation to the nature of the adrenal mass (pheochromocytoma versus adrenocortical carcinoma); the identification of cases with massive steroid excess requiring specific treatments; the selection of patients with negative prognostic biomarkers (i.e., cortisol hypersecretion) [3].

1.3   Staging and Risk Assessment

The most important prognostic factors in early ACC are disease stage, margin-free resection, age, proliferation marker Ki67, and glucocorticoid excess [5].

Tumor staging is an independent predictor of disease recurrence. Specifically, the presence of metastases is by far the strongest indicator of poor prognosis [6]. In the assessment of disease stage, guidelines recommend the tumor, node, metastasis (TNM) classification proposed by the European Network for the Study of Adrenal Tumors (ENSAT) because this system seems to be superior to others and is adopted by the Union for International Cancer Control (UICC) and World Health Organization (WHO) [3]. According to the ENSAT staging system (Table 1.2), ACC can be classified into 4 groups: stage I (≤5 cm) and stage II (>5 cm) tumors are confined to the adrenal gland; stage III tumors are extended into surrounding tissues

Table 1.2 European Network for the Study of Adrenal Tumors (ENSAT) staging system for adrenocortical carcinomas

ENSAT stage	Definition
I	T1, N0, M0
II	T2, N0, M0
III	T1–T2, N1, M0
	T3–T4, N0–N1, M0
IV	T1–T4, N0–N1, M1

T1 tumor ≤5 cm, *T2* tumor >5 cm, *T3* tumor infiltration into surrounding tissue, *T4* tumor invasion into adjacent organs or venous tumor thrombus in vena cava or renal vein. *N0* no positive lymph nodes, *N1* presence of positive lymph nodes. *M0* no distant metastases, *M1* presence of distant metastases

(i.e., para-adrenal adipose tissue or adjacent organs) or locoregional lymph nodes; stage IV means that distant metastases are present [3, 6]. In a study from the German ACC registry including 416 ACC patients, the 5-year disease-specific survival rate was 82% for patients with stage I, 58% for stage II, 55% for stage III, and 18% for stage IV ACC patients [7, 8].

Moreover, to better prognosticate patients with advanced disease, a modified ENSAT stage has been proposed to subclassify patients (mENSAT) [9]. In particular, in this modified classification, stage III includes tumors with invasion of surrounding tissues/organs or the renal/cava vein and stage IV is divided on the basis of number of metastatic organs into IVa, IVb, IVc (2, 3 or >3 metastatic organs, including N, respectively). Libé et al. demonstrated the prognostic value of this subclassification with a 5-year overall survival (OS) of 50%, 15%, 14% and 2% for stages III, IVa, IVb, and IVc, respectively [9].

Margin-free resections (R0) correlate with longer OS and recurrence-free survival (RFS) compared to patients with positive margins (R1) [5]. In the case of R0 resection, in fact, 50.4% of patients are still alive 5 years after surgery whereas the survival rate drops to 23.2% and 10.8% in R1 (not microscopically radical) or R2 (not macroscopically radical), respectively [10]. These data underline the importance of carrying out a radical surgical treatment, with en bloc removal of the tumor with clear margins in all those patients for whom the surgical option is indicated [3, 6].

Age is another independent prognostic factor; older adults usually have a poorer prognosis. This is likely multifactorial and related to increased comorbidities and reduced tolerance to systemic therapy. It is unknown if age by itself is associated with a more aggressive tumor [8]. A retrospective study that included 1579 patients found that the mortality relative risk was 1.51 in patients >55 years old compared with younger ones (95% CI 1.34–1.70) [11].

The Ki67 proliferation index is among the most important prognostic markers in ACC. The largest study, from 2015, looked at 319 German patients and 240 patients from three other European countries and showed that the hazard ratio (HR) of the RFS increased sequentially with the Ki67 index, with 10% and 20% percentage scores correlating with HRs of 1.94 (P = 0.0034) and 2.58 (P = 0.001), respectively [12]. The Ki67 index also correlates with median OS: percentage scores less than

10%, 10–19%, and ≥20% were associated with a median OS of 180.5 months, 113.5 months, and 42 months, respectively [8, 13].

Signs of cortisol excess are prognostically relevant either in terms of RFS or in terms of OS [14]. A meta-analysis of 19 studies found that the relative risk of mortality was 1.54 in hormonally functional tumors compared with hormonally non-functional tumors (95% CI 1.28–1.85) and 1.71 in cortisol-secreting tumors compared with non-cortisol secreting tumors (95% CI 1.18–2.47) [8, 15]. Among all types of hormone-secreting tumors, glucocorticoid tumors have the poorest prognosis likely due to the immunosuppressive nature and systemic effects of glucocorticoids. In a recent study, Landwehr et al. found an inverse correlation between excess glucocorticoids and tumor-infiltrating lymphocytes (TILs). The study concluded that patients with excess glucocorticoids and low numbers of TILs had a particularly poor median OS of 27 months, whereas those with sufficient numbers of TILs and no excess glucocorticoids had a median OS of 121 months [8, 16].

At the time of recurrence, the prognostic impact of disease-free interval as well as R0 status was reported in several studies [5].

In patients with metastatic disease the prognosis is generally poor but it is more heterogeneous than previously believed and long-term survivors exist. High tumor burden, high tumor grade, high Ki67 index, and uncontrolled symptoms are major factors associated with worse prognosis in these patients [6].

Different multiparametric scores have been studied in order to differentiate ACC cases on a prognostic basis and, in particular, the GRAS score has been developed. GRAS components are: grading (G, Weiss score >6 and/or Ki67 ≥20%); resection status (R), age (A) and tumor or hormone-related symptoms (S). Its prognostic value was demonstrated first in 444 patients with advanced ACC, defined as stage III or synchronous stage IV disease [9]. In particular, this study confirmed the prognostic impact of the different factors. Moreover, when the GRAS parameters were combined with the mENSAT classification, they were found to best stratify the prognosis of patients with advanced ACC. For example, 5-year OS of stage III patients ranged between 60% and 70% in patients <50 years old with an incidentally discovered ACC or with an R0 status and favorable tumor grading but they dropped to 22% when the tumor grade and the R status were both found to be unfavorable. Five-year OS for patients with stage IVa disease was 15% but ranged from 0% to 55% in patients with favorable or unfavorable GRAS parameters [9].

More recently, a modified form of the GRAS classification, termed mGRAS was proposed, which includes the ENSAT stage, focuses on Ki67 for grading and scores each parameter (Table 1.3). This modified score allows better stratification than individual clinical/histopathological characteristics identifying four subgroups with different clinical outcomes, from a more favorable prognosis (median progression-free survival [PFS] of 54 months) to a worse one (median PFS of 3 months) [17]. These data were confirmed in a large multicenter study which included 942 ACC patients who underwent surgical treatment [18]. However, further studies are needed to confirm the prognostic value of mGRAS in patients with advanced tumors.

Table 1.3 mGRAS score

mGRAS components	Group	Points
ENSAT Stage	I–II	0
	III	1
	IV	2
Grading (Ki67 index)	0–9%	0
	10–19%	1
	≥20%	2
Resection status	R0	0
	RX	1
	R1	2
	R2	3
Age	<50 years	0
	≥50 years	1
Symptoms	No	0
	Yes	1

Multiparametric scores are important as they could improve the management of ACC, personalizing the frequency of radiological surveillance, rationalizing the use of adjuvant mitotane after radical surgery and creating tailored strategies for each patient.

References

1. Terzolo M, Daffara F, Ardito A, et al. Management of adrenal cancer: a 2013 update. J Endocrinol Invest. 2014;37(3):207–17.
2. Shariq OA, McKenzie TJ. Adrenocortical carcinoma: current state of the art, ongoing controversies, and future directions in diagnosis and treatment. Ther Adv Chronic Dis. 2021;12:20406223211033103.
3. Fassnacht M, Assie G, Baudin E, et al. Adrenocortical carcinomas and malignant phaeochromocytomas: ESMO-EURACAN Clinical Practice Guidelines for diagnosis, treatment and follow-up. Ann Oncol. 2020;31(11):1476–90. Erratum in: Ann Oncol. 2023;34(7):631.
4. Ahmed AA, Thomas AJ, Ganeshan DM, et al. Adrenal cortical carcinoma: pathology, genomics, prognosis, imaging features, and mimics with impact on management. Abdom Radiol (NY). 2020;45(4):945–63.
5. Baudin E, Endocrine Tumor Board of Gustave Roussy. Adrenocortical carcinoma. Endocrinol Metab Clin North Am. 2015;44(2):411–34. Erratum in: Endocrinol Metab Clin North Am. 2015;44(3):xix.
6. Fassnacht M, Dekkers OM, Else T, et al. European Society of Endocrinology Clinical Practice Guidelines on the management of adrenocortical carcinoma in adults, in collaboration with the European Network for the Study of Adrenal Tumors. Eur J Endocrinol. 2018;179(4):G1–G46.
7. Fassnacht M, Johanssen S, Quinkler M, et al. Limited prognostic value of the 2004 International Union Against Cancer staging classification for adrenocortical carcinoma: proposal for a Revised TNM Classification. Cancer. 2009;115(2):243–50.
8. Al-Ward R, Zsembery C, Habra MA. Adjuvant therapy in adrenocortical carcinoma: prognostic factors and treatment options. Endocr Oncol. 2022;2(1):R90–R101.
9. Libé R, Borget I, Ronchi CL, et al. Prognostic factors in stage III–IV adrenocortical carcinomas (ACC): an European Network for the Study of Adrenal Tumor (ENSAT) study. Ann Oncol. 2015;26(10):2119–25.

10. Bilimoria KY, Shen WT, Elaraj D, et al. Adrenocortical carcinoma in the United States: treatment utilization and prognostic factors. Cancer. 2008;113(11):3130–6.
11. Asare EA, Wang TS, Winchester DP, et al. A novel staging system for adrenocortical carcinoma better predicts survival in patients with stage I/II disease. Surgery. 2014;156(6):1378–85; discussion 1385–6.
12. Beuschlein F, Weigel J, Saeger W, et al. Major prognostic role of Ki67 in localized adrenocortical carcinoma after complete resection. J Clin Endocrinol Metab. 2015;100(3):841–9.
13. Martins-Filho SN, Almeida MQ, Soares I, et al. Clinical impact of pathological features including the Ki-67 labeling index on diagnosis and prognosis of adult and pediatric adrenocortical tumors. Endocr Pathol. 2021;32(2):288–300.
14. Berruti A, Fassnacht M, Haak H, et al. Prognostic role of overt hypercortisolism in completely operated patients with adrenocortical cancer. Eur Urol. 2014;65(4):832–8.
15. Vanbrabant T, Fassnacht M, Assie G, Dekkers OM. Influence of hormonal functional status on survival in adrenocortical carcinoma: systematic review and meta-analysis. Eur J Endocrinol. 2018;179(6):429–36.
16. Landwehr LS, Altieri B, Schreiner J, et al. Interplay between glucocorticoids and tumor-infiltrating lymphocytes on the prognosis of adrenocortical carcinoma. J Immunother Cancer. 2020;8(1):e000469.
17. Lippert J, Appenzeller S, Liang R, et al. Targeted molecular analysis in adrenocortical carcinomas: a strategy toward improved personalized prognostication. J Clin Endocrinol Metab. 2018;103(12):4511–23.
18. Elhassan YS, Altieri B, Berhane S, et al. S-GRAS score for prognostic classification of adrenocortical carcinoma: an international, multicenter ENSAT study. Eur J Endocrinol. 2021;186(1):25–36.

Epidemiology, Presentation, Staging, and Prognostic Factors in Malignant Pheochromocytoma

2

Mara Giacché and Maria Chiara Tacchetti

2.1 Introduction

Pheochromocytomas (PHEOs) are rare neuroendocrine tumors arising from the chromaffin cells of the adrenal medulla; those arising from extra-adrenal chromaffin cells are defined paragangliomas (PGLs). The acronym PPGL comprehends both PHEO and PGL. PPGL can synthesize and secrete catecholamines (epinephrine and norepinephrine) responsible for the associated clinical syndrome. PPGL prevalence varies from 0.2% to 0.6% in hypertensive patients, to less than 0.05% in the general population. PPGLs have approximately a 15–20% 10-year probability of recurrence and a 15–20% probability of developing metastatic disease [1]. Metastatic PPGL is defined by the presence or recurrence of metastatic lesions at sites where chromaffin tissue is normally absent. Metastases can appear even 20 years after the first diagnosis: the most common sites are locoregional lymph nodes, bone (50%), liver (50%) and lung (30%) [2]. The median time for metastasis discovery is about 5 years regardless of the stage [3].

2.2 Presentation

The clinical manifestations of these tumors are primarily related to the excessive secretion of catecholamines; the amount of circulating catecholamine and the different release patterns (paroxysmal, continuous or mixed patterns) account for the variability in clinical presentation. Headache, hyperhidrosis, and palpitations constitute the classic symptomatic triad associated with PPGL. However, patients may present with many other symptoms and signs: high blood pressure, headache,

M. Giacché (✉) · M. C. Tacchetti
Internal Medicine, Department of Clinical and Experimental Sciences, University of Brescia at ASST Spedali Civili di Brescia, Brescia, Italy
e-mail: mara.giacche@asst-spedalicivili.it; mchiara.tacchetti@gmail.com

© The Author(s) 2025
G. A. M. Tiberio (ed.), *Primary Adrenal Malignancies*, Updates in Surgery,
https://doi.org/10.1007/978-3-031-62301-1_2

diaphoresis, tremors, pallor, facial flushing, shortness of breath, panic attack-type symptoms, dizziness, fatigue. Symptoms are typically paroxysmal and associated with paroxysmal hypertension. The frequency of paroxysmal attacks is highly variable: some patients experience paroxysmal episodes several times a day, others only every few months. Acute onset of symptoms may be triggered by exercise, increase in abdominal pressure, large meals, alcohol and medications (corticosteroids, ephedrine, phenylephrine, ACTH, phenothiazines, amphetamine, metoclopramide, antidepressants, some anesthetics). A variable proportion of patients (10–60% in different series) experience dizziness and faintness; these symptoms are an expression of orthostatic hypotension due to adrenergic receptor desensitization and intravascular volume depletion. About 10% of patients with secreting PPGL are normotensive, and usually reach the diagnosis either "incidentally" or thanks to application of surveillance programs in individuals carrying mutations in susceptibility genes.

Rare, serious complications of catecholamine hypersecretion are catecholamine-induced cardiomyopathies (CICMPs), which have a prevalence of 8–11% in PPGL [4]. The harmful effects of catecholamines on myocardial tissue give rise to several types of cardiomyopathies: dilated, hypertrophic, and Takotsubo. Regardless of the type of cardiomyopathy, the onset is often dramatic with acute heart failure or acute coronary syndrome. Severe hemodynamic impairment may evolve into a pheochromocytoma multisystem crisis, a rare complication with high mortality (15%), characterized by severe and prolonged hypotension with rapid progression to shock. Fortunately, with appropriate treatment these forms of cardiomyopathy are often reversible: perioperative management, surgery timing and anesthesiologic assistance must be carefully scheduled in these patients.

A less known complication of prolonged hypersecretion of catecholamines is severe constipation, which occurs in 6–7% of PPGL patients; in subjects with primary large tumors or bulky metastatic disease pseudo-obstruction may progress to paralytic ileus, bowel ischemia, and colonic perforation. Severe constipation, like catecholamine-induced cardiomyopathy, should be considered an important clue for perioperative risk stratification.

2.3   Biochemical Diagnosis

Patients with symptoms and signs compatible with catecholaminergic hyperincretion should be screened whether or not they are hypertensive. PPGL should also be excluded in subjects with changes in blood pressure during anesthesia or surgical intervention, in subjects with catecholamine-induced cardiomyopathy, in patients with adrenal incidentaloma (also if they are normotensive), in young lean individuals with diabetes mellitus type 2 (also if they do not have signs/symptoms of catecholamine excess) and in carriers of germline mutations in PPGL susceptibility genes [5].

Catecholamine excess is screened by biochemical tests: plasma or urinary free metanephrines are the most reliable indicator of tumor metabolism of catecholamines, superior to the assay of free catecholamines which are rather the expression of a secretory, often paroxysmal activity.

Plasma metanephrines and urinary free metanephrines have higher and comparable sensitivity (99% and 97%, respectively) [6]. Plasma 3-methoxytyramine can be used to detect rare dopamine-producing tumors and could also be a useful biomarker to assess the risk of malignancy. Plasmatic biochemical tests are associated with a high rate of false positive results: this is usually due to high sympathetic activity during blood sampling. To overcome this procedural error, blood sampling should be performed in a quiet room, after at least 20–30 min of supine rest; if the procedure cannot be performed with adequate accuracy, it is better to omit the plasma assay and rely on the urine test alone. A twofold increase of the upper cut-off values in one plasma metabolite or any increase in two or more metabolites have a high positive predictive value, and the patient should be referred for imaging studies [5]. Drug or food interference is responsible for false positive results: tricyclic antidepressants, α-blockers, cocaine, levodopa, MAO inhibitors, sympathomimetics, sulfasalazine may cause increased catecholamine metabolites. Caffeine, tea, alcohol, cheese, bananas, almonds, hazelnuts, vanilla should be discontinued 3 days before blood or urine sampling. Plasma chromogranin is also recommended in subjects with a clinical probability for PPGL (incidentaloma, genetic risk) once the plasma free metanephrines, 3-methoxytyramine and urinary metabolites are negative [5].

2.4   Perioperative Management

Perioperative management requires optimization of blood pressure, heart rate control and restoration of volume depletion. The alpha-adrenergic blockers doxazosin or phenoxybenzamine are traditionally considered the treatment of choice and should be given also to normotensive patients if biochemistry is indicative of catecholamine secretion. Doxazosin is a selective and competitive α1 adrenergic blocker that is given at a dose ranging from 2 to 32 mg/day in three times. Phenoxybenzamine, a non-selective and non-competitive α1–α2 adrenergic blocker, is given at a standard dose of 10 mg twice daily and can be titrated as necessary; it is not available in Italy but it is commonly prescribed in northern Europe and America. Alpha-blockers should be started at least 7–14 days before surgery [5, 7]. This recommendation has been critically questioned as it is only based on observational studies without solid evidence that the therapy with α-blockers confers any advantage in reducing perioperative mortality [8, 9]. This criticism has not received consensus and the guidelines confirm the indication for preoperative therapy with α-blockers [5, 10]. Beta-adrenergic receptor blockers can be added to control the heart rate, but only after at least 2 days of α-adrenoceptor blockade, to prevent a hypertensive crisis due to unopposed α-adrenergic receptor vasoconstriction when beta-adrenergic receptor-mediated vasodilation is blunted.

If blood pressure control is suboptimal, a calcium antagonist or renin-angiotensin blocker system can be added. Target blood pressure is lower than 130/80 mmHg in a sitting position, with a systolic blood pressure not lower than 90 mmHg while standing. A high sodium diet (5 g/day) and generous fluid intake (2.5 L/day) should be encouraged in the week before surgery; for patients with labile blood pressure values and hemodynamic instability, intravenous fluid replacement the day before surgery should be suggested.

PPGL surgery has a high risk of in intraoperative hemodynamic lability; induction of anesthesia, intubation, insufflation of peritoneum, tumor manipulation, can all trigger massive catecholamine release, so the presence of an experienced anesthesiologist is needed.

Postoperative hypotension should be treated with generous intravenous fluid replacement; also prolonged and severe hypoglycemia may appear after surgery, especially after resection of large secreting tumors: these conditions may be incorrectly diagnosed as expressions of hypoadrenalism, since after unilateral adrenalectomy adrenal function is preserved. Steroid replacement therapy is required after extensive surgery resulting in bilateral adrenalectomy or sometimes after unilateral adrenalectomy and adrenal sparing surgery on the contralateral gland.

2.5   Staging

The American Joint Committee on Cancer (AJCC) established the tumor-nodes-metastasis classification (TNM) (Table 2.1). The size of the primary tumor is a clinical predictor of metastasis, based on studies of survival and ability to metastasize; the cut-off of 5 cm was chosen to identify the transition between category T1 and T2.

Table 2.1 AJCC TNM and stage definitions for pheochromocytoma and paraganglioma

Primary tumor (T)			
TX	Primary tumor cannot be assessed		
T1	PHEO <5 cm in greatest dimension, no extra-adrenal invasion		
T2	PHEO ≥5 cm or sympathetic PGL of any size, no extra-adrenal invasion		
T3	Tumor of any size with invasion into surrounding tissues		
Regional lymph nodes (N)			
NX	Regional lymph nodes cannot be assessed		
N0	No regional lymph node metastasis		
N1	Regional lymph node metastasis		
Distant metastasis (M)			
M0	No distant metastasis		
M1	Distant metastasis		
	M1a	Metastasis to bone only	
	M1b	Metastasis to non-regional lymph node(s), liver and/or lung; no skeletal metastasis	
	M1c	Metastasis to bone and multiple other sites	
AJCC prognostic stage groups			
Stage I	T1	N0	M0
Stage II	T2	N0	M0
Stage III	T1–2	N1	M0
	T3	N0–1	M0
Stage IV	T1–3	N0–1	M1

AJCC American Joint Committee on Cancer, *PHEO* pheochromocytoma, *PGL* paraganglioma

TNM and AJCC prognostic stage groups have been shown to correlate with overall survival. Stage I includes patients who have very low metastatic potential and excellent prognosis, with the exception of patients with mutant succinate dehydrogenase subunit B gene (*SDHB*) for whom, even in the presence of tumor <5 cm without invasion of surrounding tissues, the risk of metastasis is higher: patients with *SDHB* germline mutation can be considered already stage II, regardless of the size of the tumor [3].

2.6 Prognostic Factors

PPGL has the potential to metastasize and is therefore defined as tumor with uncertain biological behavior. When not metastatic at onset, distinguishing a benign PPGL from tumors with metastatic potential is very challenging. Several anatomo-pathological, molecular and biological characteristics have been identified as possible prognostic factors of malignancy. A Ki67 index greater than 3% is a reliable indicator of proliferating cells and predictor of tumor progression (high specificity); however, it is often low or negative in PPGL (low sensitivity). It has been well demonstrated that norepinephrine-secreting tumors (predominantly paragangliomas) are more frequently malignant than those secreting epinephrine (almost exclusively PHEOs), and high plasmatic 3-methoxytyramine correlates with poorly differentiated PPGL.

The Pheochromocytoma of the Adrenal gland Scaled Score (PASS) is a scoring system which involves multiple histological features (Table 2.2) [11]. A PASS ≥4 is suggestive for PPGL with metastatic potential, with 50% sensitivity and 45% specificity. Unfortunately, the reliability of the PASS is affected by the subjective interpretation of the pathologist, limiting its clinical utility.

The GAPP (Grading system for Adrenal Pheochromocytoma and Paraganglioma) scoring system includes histopathologic characteristics and biochemical profile (Table 2.3) [11]. According to the GAPP score, PPGLs are classified as well,

Table 2.2 PASS (Pheochromocytoma of the Adrenal gland Scaled Score)

PASS parameters	Points
Nuclear hyperchromasia	1
Profound nuclear pleomorphism	1
Capsular invasion	1
Vascular invasion	1
Extension into periadrenal adipose tissue	2
Atypical mitotic figures	2
>3 mitotic figures/10 HPF	2
Tumor cell spindling	2
Cellular monotony	2
High cellularity	2
Central or confluent tumor necrosis	2
Large nests or diffuse growth (>10% of tumor volume)	2
Total maximum score	20

Scores ≥4 are predictive of malignancy

Table 2.3 GAPP (Grading system for Adrenal Pheochromocytoma and Paraganglioma) score

GAPP parameters	Points
Histological pattern	
Zellballen	0
Large and irregular nest	1
Pseudorosette (even focal)	1
Comedo-type necrosis	
Absence	0
Presence	2
Cellularity	
Low (<150 cells/U)	0
Moderate (150–250 cells/U)	1
High (> 250 cells/U)	2
Ki67 labeling index (%)	
<1	0
1–3	1
>3	2
Vascular or capsular invasion	
Absence	0
Presence	2
Catecholamine type	
Non-functioning	0
Adrenergic type	0
Noradrenergic type	1
Total maximum score	10

According to their GAPP scores, tumors are classi-fied as: well differentiated (0–2); moderately differentiated (3–6); poorly differentiated (7–10)

moderately, and poorly differentiated, and this classification correlates with 10-year survival rates (83%, 38%, and 0%, respectively).

The analysis of gene expression in tumor tissue has highlighted how detection of a larger number of somatic mutations is associated with worse outcome. The presence of a germline mutation of *SDHB* gene is a well-known predictor of malignancy. About 50% of *SDHB*-mutated patients have a metastatic disease at onset or develop metastasis during the follow-up. The presence of *SDHB* mutation is also associated with reduced median overall survival (42 vs. 244 months in non-*SDHB* mutant metastatic PPGL) [11]. An up-to-date GAPP score (M-GAPP) includes loss of *SDHB* immunohistochemistry staining as a surrogate for *SDHB* expression but it has yet to be validated (Table 2.4).

Recent molecular studies have identified three different molecular signatures in PPGL; in accordance with the germinal and/or somatic mutation leading to the activation of oncogenic signaling pathway, the tumor is attributed to a specific cluster: the pseudo-hypoxic cluster, the kinase cluster, and the Wnt cluster. These three molecular clusters differ in phenotype and clinical behavior. Somatic or germinal mutations in the Krebs cycle-associated genes (*SDHx, FH, MDH2, GOT2, SLC25A11, DLST, IDH1*) and VHL/EPAS1 related genes (*PHD1/2, EGLN1, EPAS1, IRP1*) cause activation of pseudohypoxic pathway (cluster 1) and are associated with a more aggressive behavior.

Table 2.4 M-GAPP score

M-GAPP parameters	Points
Histological pattern	
Zellballen	0
Large and irregular nest or pseudorosette	2
Comedo-type necrosis	
Absence	0
Presence	2
Ki67 labeling index (%)	
<1	0
≥1	2
Vascular or capsular invasion	
Absence	0
Presence	1
Catecholamine type	
Non-functioning	0
Adrenergic type	0
Noradrenergic type	1
SDHB immunochemistry	
Positive	0
Negative	2
Total maximum score	10

Scores ≥3 are predictive of malignancy

2.7   Postoperative Follow-Up

Patients operated on for PPGL need a follow-up of at least 10 years to screen for local or metastatic recurrences. Patients with high risk of recurrence or malignancy (genetic risk, very large tumors, unfavorable prognostic factors) should be offered lifelong follow-up. Plasma and/or urinary free metanephrines should be tested annually; subjects with normal preoperative levels of metanephrines and elevated chromogranin-A should be screened with annual chromogranin-A assessment. For patients with completely negative preoperative biochemistry, follow-up should be performed with imaging examinations every 1–2 years [12].

References

1. Granberg D, Juhlin CC, Falhammar H. Metastatic pheochromocytomas and abdominal paragangliomas. J Clin Endocrinol Metab. 2021;106(5):e1937–52.
2. Angelousi A, Kassi E, Zografos G, Kaltsas G. Metastatic pheochromocytoma and paraganglioma. Eur J Clin Invest. 2015;45(9):986–97.
3. Jimenez C, Ma J, Roman Gonzalez A, et al. TNM Staging and overall survival in patients with pheochromocytoma and sympathetic paraganglioma. J Clin Endocrinol Metab. 2023;108(5):1132–42.
4. Kumar A, Pappachan JM, Fernandez CJ. Catecholamine-induced cardiomyopathy: an endocrinologist's perspective. Rev Cardiovasc Med. 2021;22(4):1215–28.
5. Lenders JWM, Kerstens MN, Amar L, et al. Genetics, diagnosis, management and future directions of research of phaeochromocytoma and paraganglioma: a position statement and

consensus of the Working Group on Endocrine Hypertension of the European Society of Hypertension. J Hypertens. 2020;38(8):1443–56.

6. Eisenhofer G, Prejbisz A, Peitzsch M, et al. Biochemical diagnosis of chromaffin cell tumors in patients at high and low risk of disease: plasma versus urinary free or deconjugated o-methylated catecholamine metabolites. Clin Chem. 2018;64(11):1646–56.

7. Lenders JW, Duh QY, Eisenhofer G, et al. Pheochromocytoma and paraganglioma: an endocrine society clinical practice guideline. J Clin Endocrinol Metab. 2014;99(6):1915–42. Erratum im: J Clin Endocrinol Metab. 2023 Apr 13;108(5):e200

8. Groeben H, Nottebaum BJ, Alesina PF, et al. Perioperative α-receptor blockade in phaeochromocytoma surgery: an observational case series. Br J Anaesth. 2017;118(2):182–9.

9. Lentschener C, Baillard C, Dousset B, Gaujoux S. Dogma is made to be broken. Why are we postponing curative surgery to administer ineffective alpha adrenoreceptor blockade in most patients undergoing pheochromocytoma removal? Endocr Pract. 2019;25(2):199.

10. Fassnacht M, Assie G, Baudin E, et al. Adrenocortical carcinomas and malignant phaeochromocytomas: ESMO-EURACAN Clinical Practice Guidelines for diagnosis, treatment and follow-up. Ann Oncol. 2020;31(11):1476–90. Erratum in: Ann Oncol. 2023;34(7):631.

11. Wachtel H, Hutchens T, Baraban E, et al. Predicting metastatic potential in pheochromocytoma and paraganglioma: a comparison of PASS and GAPP scoring systems. J Clin Endocrinol Metab. 2020;105(12):e4661–70.

12. Plouin PF, Amar L, Dekkers OM, et al. European Society of Endocrinology Clinical Practice Guideline for long-term follow-up of patients operated on for a phaeochromocytoma or a paraganglioma. Eur J Endocrinol. 2016;174(5):G1–10.

Genetics and Molecular Biology of Adrenocortical Carcinoma

3

Salvatore Grisanti, Chiara Romani, Marta Laganà, and Deborah Cosentini

3.1   Introduction

In the last decade, integrated multi-omic platforms revealed that adrenocortical carcinoma (ACC) is characterized by a complex genomic organization (germline and somatic DNA genes, aneuploidy, DNA proliferation and translation, epigenetics), and complex proteomic and metabolomic profiles. This pathological complexity can be seen as the result of dysregulation of the tight controls that regulate the development and the multiple functions of the normal adrenal gland [1]. There are several clinical consequences of this biological complexity:

- ACC is not a monogenic but rather a polygenic disease;
- one therapeutic target does not fit all ACC heterogeneity;
- ACC is an adaptive disease characterized by temporal heterogeneity.

S. Grisanti (✉) · M. Laganà · D. Cosentini
Medical Oncology, Department of Medical and Surgical Specialties, Radiological Sciences, and Public Health, University of Brescia at ASST Spedali Civili di Brescia, Brescia, Italy
e-mail: salvatore.grisanti@unibs.it; marta.lagana@unibs.it; deborah.cosentini@unibs.it

C. Romani
Department of Medical and Surgical Specialties, Radiological Sciences, and Public Health, University of Brescia, Brescia, Italy
Angelo Nocivelli Institute for Molecular Medicine at ASST Spedali Civili di Brescia, Brescia, Italy
e-mail: chiara.romani@unibs.it

© The Author(s) 2025
G. A. M. Tiberio (ed.), *Primary Adrenal Malignancies*, Updates in Surgery,
https://doi.org/10.1007/978-3-031-62301-1_3

3.2   Germline DNA Mutations and the Hereditary Component of Adrenocortical Carcinoma

The heritable fraction of ACC is estimated at 5–10% and 50–80% of adult and pediatric cases, respectively [2]. ACC is a rare cancer and population-based registries of patients with hereditary ACC are lacking worldwide except in Southern Brazil where there is abundance of specific germline mutations of the *TP53* gene causing a higher incidence of ACC [3]. Elsewhere, ACC can arise in the context of cancer-predisposing syndromes: Li-Fraumeni syndrome (*TP53*), Lynch syndrome (*NMR* genes), Beckwith-Wiedemann syndrome (*CDKN1C, H19, IGF2, KCNQ1OT1*), Carney complex (*PRKAR1A*), multiple endocrine neoplasia type 1 (*MEN1*) [2]. Other rarer germinal variants potentially predisposing to ACC have been described in succinate dehydrogenase genes (*SDHx*) [4] and in Armadillo-containing repeat protein 5 gene (*ARMC5*) [5]. In the majority of adult cases, however, ACC is diagnosed as a sporadic cancer with acquired genomic alterations of the somatic DNA.

The Cancer Genome Atlas (TCGA) project analyzed germline variants (GVs) relevant to adult ACC in a pan-cancer study and in a specific ACC study. From the analysis of the core dataset of 91 ACC cases in the TCGA pan-cancer study [6], a low rate of GVs was found that places adult ACC in the lowest quartile among the 33 cancers screened. In the TCGA-ACC study [7], 9 GVs were found among 177 genes potentially linked to ACC. In a recent Italian study, 21 GVs among 17 genes (including *TP53, ARMC5* and DNA Damage Repair [DDR] genes) were found in 150 (9.3%) patients with sporadic ACC that correlated with shorter survival [8].

There are practical implications concerning germline DNA in ACC. First, because ACC is not included in any screening program of hereditary cancer syndromes (like colorectal, breast or thyroid cancers), patients with a hereditary cancer syndrome should undergo surveillance for ACC. Second, patients with a diagnosis of ACC should be tested to identify germline mutations in cancer-predisposing genes in their families [9]. Third, GVs affecting DDR genes could have therapeutic and prognostic relevance.

3.3   Chromosomal Number Alteration (Aneuploidy)

Whole chromosome or chromosome arm imbalance is called aneuploidy, a hallmark of human cancers [10]. Chromosome copy number aberrations (CNA) include whole chromosome or single-arm alterations as well as smaller changes, like loss of heterozygosity (LOH), or larger changes, like whole genome doubling (WGD). Higher tumor aneuploidy negatively correlates with T cell infiltration, T cell clonality, expression of immune-related genes and overall survival [11].

ACC is frequently hypodiploid compared with other cancer types. However, copy number gains and losses can occur in up to 60% of cases (a pattern called "noisy" in the TCGA-ACC study). This pattern of high aneuploidy is often associated with WGD, which is related to alteration of the telomeres' length regulation machinery leading to cell immortalization and is a clinical marker of poor prognosis [7].

3.4   Somatic DNA Mutations and Tumor Mutation Burden

Several somatic DNA gene mutations and corresponding functional pathways have been found in ACC. Most of our current knowledge of ACC genomics derives from three main multi-omics studies of structural and functional alterations in this disease [7, 12, 13]. While the list of single somatic gene mutations is constantly increasing, there are recurring mutations (frequency >10% indicated in square brackets) of candidate driver genes in ACC that are summarized as follows:

- genes involved in cell cycle (*TP53* [21%], *CDKN2A* [15%], *RB1*, *CDK4*, *CCNE1*);
- genes involved in Wnt/β-catenin signaling (*CTNNB1* [16%], *ZNRF3* [19%]);
- genes involved in chromatin remodeling (*MEN1*, *DAXX*);
- genes involved in telomere maintenance (*TERT* [14%], *TERF2*);
- genes involved in protein kinase cAMP-dependent regulatory type I alpha (cAMP/PKA signaling) (*PRKAR1A* [11%]) [14];
- genes involved in DNA transcription (*MED12*) and RNA translation (*RPL22*).

An exception is represented by the alteration of the insulin-like growth factor-2 gene (*IGF2*) that is a hallmark of ACC (loss of heterozygosity in 90% of cases) [15, 16] but overexpression of the corresponding insulin-like growth factor-2/receptor-1 (IGF2/IGF1R) axis seems not to be a driver pathway in ACC, as demonstrated by failure of a phase III clinical intervention with the anti-IGF2/IGF1R drug linsitinib [17].

Collectively, the two most frequently altered genes and functional pathways in ACC are the TP53/RB1 cell cycle and the Wnt/β-catenin pathways (33–45% and 41% of cases, respectively) [7, 12, 13].

- *DNA damage repair (DDR) genes*
  In the TCGA pan-cancer study, >80% of ACC samples displayed at least one DDR gene alteration including genes involved in:
  - mismatch repair (MMR): *MLH1–3*, *MSH2–6*, *PMS2*;
  - homologous recombination (HR): *TP53BP1*, *BRCA1–2*, *BRIP1*, *RAD51*, *TOP3A*;
  - damage sensor (DS): *ATM*, *ATR*, *CHEK2*;
  - translesion synthesis (TS): *REV3L*;
  - base excision repair (BER): *POLB*.

  Other minor DDR gene alterations involve direct repair (*ALKBH3*, *MGMT*), Fanconi anemia (*FANCA*, *FANCD2*) and nonhomologous end joining (*LIG4*, *XRCC4*, *XRCC6*) [18].
  Many of these gene alterations are found at the level of both germinal and somatic DNA: in particular, germinal MMR gene alterations observed in familial ACC cases identify familial ACC as a Lynch syndrome (LS)-associated cancer [19].
- *Microsatellite instability (MSI)*
  Defects of the MMR system cause microsatellite instability (MSI) that is both prognostic and predictive of response to therapy in many cancer types including

colorectal and endometrial cancers. In a pan-cancer re-analysis of TCGA data, Bonneville et al. found a MSI-high (MSI-H) phenotype in 4.3% of ACC cases placing ACC as the fifth neoplasm with the highest MSI-H rate among 39 different cancers [20].

* *Tumor mutation burden (TMB)*
  In the TCGA-ACC study, the median somatic mutation density was 0.9 mutation/Mb (range 0.2–14.0 mutations/Mb) [7]. In a pan-cancer analysis, ACC had a median TMB less than 5 mutations/Mb and less than 10% of cases had a TMB >10 mutations/Mb. Therefore, despite all of the above considerations, ACC is placed among tumors with the lowest TMB [21].

From the above considerations it does appear clear that ACC is not driven by one single gene alteration (like the GIST with *c-Kit*). A mosaic of multiple, concurrent and not exclusive gene alterations can be present, generating a high degree of pleiotropism.

Thus, translation of genomic data into effective target therapies has been hampered because of two main reasons. First, many gene alterations in ACC are master regulators of fundamental processes in eukaryotic cells (e.g., TP53/Rb or Wnt/β-catenin) for which there are no specific drugs, or they could cause unacceptable toxicities. Second, while one biological pathway may be predominant in ACC, there are multiple ways to escape or circumvent that pathway.

3.5   Epigenetic (Post-Translational) Changes

In biology, epigenetics refers to changes of gene expression without changes of the DNA sequence that occur by activation or repression of specific transcription factors or biochemical modifications (acetylation, methylation) of specific target genes. At least seven studies identified DNA methylation as an important mechanism of epigenetic control of gene expression in ACC. Both hypomethylation and hypermethylation of promoter regions can occur at a higher frequency in ACC compared to adrenal adenomas [22, 23]. In the European Network for the Study of Adrenal Tumors (ENSAT) and TCGA-ACC studies, analysis of the hypermethylation at CpG-rich islands defined three clusters of methylation (CpG island methylator phenotype [CIMP]-high, -intermediate, and -low) that showed a significant prognostic value. In particular, the CIMP-high profile identified ACC patients with higher proliferative index and worse prognosis [7, 12]. For further reading, a comprehensive review by Ettaieb et al. on the role of epigenetic alterations in ACC and their potential role as prognostic factors and therapeutic targets has been published [24].

3.6   An Integrative View of Molecular Biology of Adrenocortical Carcinoma

The availability of multi-omics platforms enabled researchers to provide an integrative view of the biology of ACC by combining germinal and somatic DNA, epigenetic changes, chromosomal aberrations, RNA, proteins and other metabolites. In particular, the TCGA-ACC study proposed a molecular classification of ACC with identification of three different clusters of diseases each characterized by homogeneous molecular subtypes in terms of DNA, RNA, proteins, etc. The final clusters were called "cluster of clusters" (COC) numbered from 1 to 3.

Patients in COC1 had better prognosis, lower grade of aneuploidy, lower somatic mutations, lower genome-wide methylation rate (CIMP-low), high expression of IGF2, low expression of steroidogenic machinery and higher immune cell infiltration signature. Patients in COC3 had the worst prognosis, frequent mutations involving the cell cycle and DNA damage repair pathways, overexpression of Wnt/β-catenin pathway, higher degree of aneuploidy and WGD, high CIMP profile, high expression of steroidogenic machinery and the lowest expression of immune cell infiltration signature. Patients in COC2 have features of intermediate prognosis and biology [7, 25].

Despite the fact that a molecular classification of ACC is not validated and, therefore, not ready yet for clinical use, this approach will match clinical and molecular profiles and it will hopefully generate new venues for a more rationale therapeutic strategy.

3.7   Spatial/Temporal Molecular Heterogeneity in Adrenocortical Carcinoma

Clinical decisions are taken on the basis of a biological snapshot of the ACC disease taken at a given time (usually at diagnosis). However, little is known about whether metastases from ACC share the same genomic alterations of their primary tumor and, more importantly, if their genotype confers different sensitivity or resistance to therapy. In a study investigating genomic heterogeneity in 33 metastatic ACCs, investigators demonstrated that ACC is characterized by a significant heterogeneity among different metastatic sites in the same patient and metastases had a 2.8-times higher mutation rate than the primary ACC [26].

References

1. Grisanti S, Cosentini D, Sigala S, Berruti A. Molecular genotyping of adrenocortical carcinoma: a systematic analysis of published literature 2019–2021. Curr Opin Oncol. 2022;34(1):19–28.
2. Else T. Association of adrenocortical carcinoma with familial cancer susceptibility syndromes. Mol Cell Endocrinol. 2012;351(1):66–70.
3. Ribeiro RC, Sandrini F, Figueiredo B, et al. An inherited p53 mutation that contributes in a tissue-specific manner to pediatric adrenal cortical carcinoma. Proc Natl Acad Sci U S A. 2001;98(16):9330–5.
4. Else T, Lerario AM, Everett J, et al. Adrenocortical carcinoma and succinate dehydrogenase gene mutations: an observational case series. Eur J Endocrinol. 2017;177(5):439–44.
5. Stratakis CA, Berthon A. Molecular mechanisms of ARMC5 mutations in adrenal pathophysiology. Curr Opin Endocr Metab Res. 2019;8:104–11.
6. Huang KL, Mashl RJ, Wu Y, et al. Pathogenic germline variants in 10,389 adult cancers. Cell. 2018;173(2):355–70.e14.
7. Zheng S, Cherniack AD, Dewal N, et al. Comprehensive pan-genomic characterization of adrenocortical carcinoma. Cancer Cell. 2016;29(5):723–36. Erratum in: Cancer Cell. 2016;30(2):363.
8. Grisanti S, Scatolini M, Tomaiuolo P, et al. Germline variants NGS characterization in patients with non-syndromic adrenocortical carcinoma. Abstract 341. ESMO Sarcoma and Rare Cancers 2023 Congress. ESMO Open. 2023;8(1 Suppl 3):101050. https://doi.org/10.1016/j.esmoop.2023.101050.
9. Else T, Rodriguez-Galindo C. 5th International ACC Symposium: Hereditary predisposition to childhood ACC and the associated molecular phenotype: 5th International ACC Symposium Session: Not just for kids! Horm Cancer. 2016;7(1):36–9.
10. Taylor AM, Shih J, Ha G, et al. Genomic and functional approaches to understanding cancer aneuploidy. Cancer Cell. 2018;33(4):676–89.e3.
11. Davoli T, Uno H, Wooten EC, Elledge SJ. Tumor aneuploidy correlates with markers of immune evasion and with reduced response to immunotherapy. Science. 2017;355(6322):eaaf8399.
12. Assié G, Letouzé E, Fassnacht M, et al. Integrated genomic characterization of adrenocortical carcinoma. Nat Genet. 2014;46(6):607–12.
13. Juhlin CC, Goh G, Healy JM, et al. Whole-exome sequencing characterizes the landscape of somatic mutations and copy number alterations in adrenocortical carcinoma. J Clin Endocrinol Metab. 2015;100(3):E493–502.
14. Ronchi CL. cAMP/protein kinase A signaling pathway and adrenocortical adenomas. Curr Opin Endocr Metab Res. 2019;8:15–21.
15. Angelousi A, Kyriakopoulos G, Nasiri-Ansari N, et al. The role of epithelial growth factors and insulin growth factors in the adrenal neoplasms. Ann Transl Med. 2018;6(12):253.
16. Altieri B, Colao A, Faggiano A. The role of insulin-like growth factor system in the adrenocortical tumors. Minerva Endocrinol. 2019;44(1):43–57.
17. Fassnacht M, Berruti A, Baudin E, et al. Linsitinib (OSI-906) versus placebo for patients with locally advanced or metastatic adrenocortical carcinoma: a double-blind, randomised, phase 3 study. Lancet Oncol. 2015;16(4):426–35.
18. Knijnenburg TA, Wang L, Zimmermann MT, et al. Genomic and molecular landscape of DNA damage repair deficiency across The Cancer Genome Atlas. Cell Rep. 2018;23(1):239–54.e6.
19. Raymond VM, Everett JN, Furtado LV, et al. Adrenocortical carcinoma is a lynch syndrome-associated cancer. J Clin Oncol. 2013;31(24):3012–8. Erratum in: J Clin Oncol. 2013;31(28):3612.
20. Bonneville R, Krook MA, Kautto EA, et al. Landscape of microsatellite instability across 39 cancer types. JCO Precis Oncol. 2017;2017:PO.17.00073.
21. Yarchoan M, Albacker LA, Hopkins AC, et al. PD-L1 expression and tumor mutational burden are independent biomarkers in most cancers. JCI Insight. 2019;4(6):e126908.

22. Rechache NS, Wang Y, Stevenson HS, et al. DNA methylation profiling identifies global methylation differences and markers of adrenocortical tumors. J Clin Endocrinol Metab. 2012;97(6):E1004–13.
23. Fonseca AL, Kugelberg J, Starker LF, et al. Comprehensive DNA methylation analysis of benign and malignant adrenocortical tumors. Genes Chromosomes Cancer. 2012;51(10):949–60.
24. Ettaieb M, Kerkhofs T, van Engeland M, Haak H. Past, present and future of epigenetics in adrenocortical carcinoma. Cancers (Basel). 2020;12(5):1218.
25. Mohan DR, Lerario AM, Hammer GD. Therapeutic targets for adrenocortical carcinoma in the genomics era. J Endocr Soc. 2018;2(11):1259–74.
26. Gara SK, Lack J, Zhang L, et al. Metastatic adrenocortical carcinoma displays higher mutation rate and tumor heterogeneity than primary tumors. Nat Commun. 2018;9(1):4172.

Genetics and Molecular Biology of Pheochromocytoma and Paraganglioma

4

Mara Giacché, Maria Chiara Tacchetti, and Maurizio Castellano

4.1 Introduction

Pheochromocytoma (PHEO) and paraganglioma (PGL) are currently considered the tumors with the greatest genetic determinism: in addition to hereditary PHEO associated with neurofibromatosis, Von Hippel Lindau syndrome and multiple endocrine neoplasia type 2 syndrome, many other susceptibility genes have been discovered, identifying PHEO and PGL (PPGL) as tumors with high genetic heterogeneity. The application of genetic screening to all PPGL patients, irrespective of family history or syndromic features, allows detection of a germline mutation in 40% of subjects, with a germline mutation frequency of about 10–12% also in patients with sporadic presentation. Predisposition to PPGL is mostly transmitted in an autosomal dominant fashion, even if large variability in penetrance does not always permit recognition of the traces of hereditability, and one or more generations appear to be skipped. Bilateral PHEO or multifocal PGL, recurrent or malignant disease along with a young age of onset (<45 years) are all possible signs of inherited disease; however, due to the high prevalence of germline mutations also in apparently sporadic disease, the application of genetic screening is currently recommended for all patients with PPGL [1, 2]. The identification of a specific hereditary form has implications for the correct management/follow-up of the patient and also allows extension of genetic analysis to relatives and implementation of presymptomatic surveillance in mutation carriers.

M. Giacché (✉) · M. C. Tacchetti
Internal Medicine, Department of Clinical and Experimental Sciences, University of Brescia at ASST Spedali Civili di Brescia, Brescia, Italy
e-mail: mara.giacche@asst-spedalicivili.it; mchiara.tacchetti@gmail.com

M. Castellano
Department of Clinical and Experimental Sciences, University of Brescia at ASST Spedali Civili di Brescia, Brescia, Italy
e-mail: maurizio.castellano@unibs.it

© The Author(s) 2025
G. A. M. Tiberio (ed.), *Primary Adrenal Malignancies*, Updates in Surgery,
https://doi.org/10.1007/978-3-031-62301-1_4

The current availability of high-throughput gene sequencing techniques (next generation sequencing, NGS) allows for the simultaneous study of many genes, overcoming the difficulties associated with the study of a disease with high genetic heterogeneity [3]. Sequencing detects small intragenic insertion or deletion, missense, no sense and splice variants, but whole gene deletion and large intragenic deletion or duplication are not detected; for this analytical purpose other methods must be used, such as quantitative C-reactive protein, multiple ligation-dependent probe amplification, or gene-targeted microarray, according to laboratory preference.

NGS panels available for the study of PPGL susceptibility genes are usually customized with the genes most solidly associated with the development of the disease; the identification of a pathologic variant in one of these genes makes it possible to propose to the patient a surveillance program according to the risk of relapse, malignancy, and involvement of other organs. Larger panels, including more recently identified genes, are available in clinical research centers, though the characteristics of penetrance and phenotype expressivity for mutations in these genes are not yet defined, and the clinical interpretation in many cases cannot be conclusive.

In this chapter, the genetics of PPGL will be described starting from the first recognized syndromic forms up to the recently identified susceptibility genes, with a focus on the specific clinical implications associated with the different genes involved (Table 4.1).

Table 4.1 Clinical phenotype of genetic pheochromocytoma/paraganglioma (PPGL) syndromes

Gene	Hereditary PPGL %[a]	Location of tumor	Median age at presentation	Presentation in childhood	Penetrance at 60–70 years	Risk of malignancy
NF1	2–3	Adrenal Bilateral (15%)	40–50	Rare	3%	10–12%
VHL	5–7	Adrenal Bilateral (50%)	18–30	40% of pediatric PHEO	20–24%	5–8%
RET	6	Adrenal Bilateral (50–70%)	30–40	Described	20–50%	<5%
SDHA (PGL 5)	5–7	Head PGL Abdominal/ thoracic PGL	35–43	Described	10%	30–66%
SDHB (PGL 4)	10	Abdominal/ thoracic PGL Head and neck PGL Multiple (<20%)	25–30	Described	25%	35–75%

Table 4.1 (continued)

Gene	Hereditary PPGL %[a]	Location of tumor	Median age at presentation	Presentation in childhood	Penetrance at 60–70 years	Risk of malignancy
SDHC (PGL3)	1	Head and neck PGL Abdominal/thoracic PGL Multiple (25–30%)	35–40	–	25%	Low
SDHD (PGL1)	9	Head and neck PGL Multiple (>50%)	30–40	Rare	43–86%	15–29%
SDHAF2 (PGL2)	<1	Head and neck PGL Multiple (>50%)	25–45	–	Probably high (>50%)[b]	Low
TMEM	1–2	Adrenal Bilateral (rare)	35–40	–	Unknown	Low
MAX	1	Adrenal Bilateral (50%)	35–40	–	Probably high[b]	<10%[b]
FH	1	Abdominal PGL	30–40	–	Unknown	30%[b]
Other genes	<1	–	–	–	–	–

PHEO pheochromocytoma, *PGL* paraganglioma
[a] Proportion of hereditary PPGL attributed to variant in gene
[b] Preliminary data

4.2 *NF1* Gene

NF1 gene is a tumor suppressor gene responsible for neurofibromatosis type 1 (NF1), an autosomal dominant disease. The prevalence of PPGL in *NF1* patients is lower than 3%. Frequently the diagnosis is incidental. In most cases the tumor is a single PHEOs, extra-adrenal PGLs are rarer. The mean age at diagnosis is 40 years, similar to sporadic disease, even though cases in young subjects (before 20 years of age) have also been described. Bilateral disease occurs in about 15% of patients, and this relevant information must be kept in mind to favor adrenal-sparing surgery. About 10–12% of NF1-PPGL are malignant.

NF1 is a large gene, and it is not routinely comprised in PPGL-NGS panels: as the penetrance of disease is virtually complete by the age of 7, an appropriate anamnesis and clinical examination are sufficient for the diagnosis.

4.3 *RET* Gene

The *RET* gene is a protooncogene, which encodes a transmembrane tyrosine kinase: gain of function mutations activate RET kinase activity conferring oncogenic properties. *RET* is the causative gene of multiple neoplasia endocrine syndrome type 2 (Men2), a disease with autosomal dominant transmission. In the Men2 syndrome, PHEO, together with medullary thyroid cancer, is one of the two main clinical manifestations, and has a penetrance of about 20–50%. In 10% of Men2, PHEO is the first clinical manifestation. It usually presents at the age of 30–40 years, but it can also develop in infancy by the age of 11. PHEO may be bilateral at presentation or may recur in the contralateral gland, about 50–70% of patients develop bilateral disease; extra-adrenal PGL are very rare, but have been described. They are very rarely malignant (fewer than 5%). Due to the high risk of recurrence, adrenal-sparing surgery is advocated.

4.4 *VHL* Gene

VHL is an oncosuppressor gene and is the causative gene of Von Hippel Lindau syndrome. This hereditary neoplastic syndrome, with autosomal dominant transmission, is characterized by multiorgan involvement; tumors which most impact on the clinical course of disease are central nervous system and retinal hemangioblastomas, clear cell carcinoma, neuroendocrine tumors and PPGL. Penetrance for PPGL is estimated to be 20–24%, sometimes with a very young age of onset (18–30 years); *VHL* gene mutations are in fact identified in about 40% of pediatric PHEOs. Tumors are mostly adrenal and are often bilateral (50%), frequently even at presentation. Malignant disease is rare (5–8%). Due to the high risk of bilateral/recurrent disease, adrenal surgery should always be preferred.

4.5 *SDHx* Genes and *SDHAF2* Gene

The *SDHx* genes (*SDHA*, *SDHB*, *SDHC*, and *SDHD*) codify for the four subunits of the succinate dehydrogenase (SDH), a mitochondrial enzyme involved in the transfer of electrons in the mitochondrial respiratory chain; SDHAF2 is a mitochondrial protein required for the activation of the SDH complex. *SDHx* and *SDHAF2* are oncosuppressor genes: mutations in these genes were identified in the 2000s as responsible for nearly 50% of hereditary PPGL. *SDHx* mutations may also predispose to kidney cancer and GIST (wild c-kit, wild PDGFRA). Some particularities for each of these genes are briefly described below.

4.5.1 *SDHD* Gene (PGL1 Syndrome)

Mutations in the *SDHD* gene are characterized by high penetrance, about 86% for the age of 50; this explains why most of the subjects with *SDHD*-related PPGL have familial antecedents [4]. The model of hereditary transmission is autosomal dominant, with maternal imprinting: this means that the phenotype is expressed mostly with paternal transmission, while only 5% of maternal inherited *SDHD* mutations express the phenotype [5]. The typical *SDHD*-related phenotype is head and neck PGL (HNPGL), even though thoracic-abdominal or pelvic PGL have been described. The mean age of onset is 36 years, and in more than half of subjects PGLs are multiple and/or bilateral. Malignancy is described in 15–29% of cases [6].

4.5.2 *SDHAF2* Gene (PGL2 Syndrome)

Mutations in the *SDHAF2* gene are a very rare cause of familial PGL. As for *SDHD* mutations, the transmission is autosomal dominant with a parent-of-origin effect, so cancer susceptibility is expressed only when the mutation is inherited from the father [7]. The usual phenotype is multiple HNPGLs, with a high penetrance [8], even though only a few families have been described and these data need to be confirmed. The mean age of onset is 33 years, and no malignancy has been described so far.

4.5.3 *SDHC* Gene (PGL3 Syndrome)

Mutations in the *SDHC* are a rare cause of hereditary PGL, transmission is autosomal dominant. Patients develop mostly HNPGLs, even though PHEO and extra-adrenal PGLs have been described. Penetrance is not defined, only 25% of patients have a suggestive family history [4], assuming a lower penetrance compared to *SDHD* mutations. Average age at diagnosis is 38 years, the risk of malignancy is low.

4.5.4 *SDHB* Gene (PGL4 Syndrome)

Mutations in *SDHB* account, together with mutations in *SDHD*, for most *SDHx*-related PPGLs. *SDHB* mutation carriers develop abdominal, pelvic, thoracic, or cervical PGLs, less frequently PHEO. The average age of onset of PPGL is 25–30 years, although the disease may appear at very young age (6–7 years). The most relevant clinical feature of the *SDHB* gene is the association with malignant disease occurring in more than 50% of *SDHB*-related PGL; it is estimated that about 36% of all malignant PPGL are due to *SDHB* mutations [9]. Penetrance is lower compared to *SDHD*/*SDHC* mutations and is estimated to be about 25–40%.

4.5.5 *SDHA* Gene (PGL5 Syndrome)

The *SDHA* gene encodes the catalytic subunit A of the SDH complex. Biallelic mutations of the *SDHA* gene are responsible for Leigh's syndrome, a severe early-onset and progressive neurometabolic disorder.

Association of the *SDHA* gene with PPGL has been more recently identified: heterozygous mutations in *SDHA* are in fact a rare cause of PPGL and have been identified in subjects with abdominal, pelvic, thoracic, thyroid or cervical PGL. The model of inheritance is autosomal dominant, penetrance is low (about 10% for the age of 70), mean age of onset is 43 years [10]. *SDHA* mutations have been associated with malignant disease in more than 30% of cases [11].

4.6   *TMEM127* Gene

TMEM127 mutations predispose to PHEO. Bilateral disease has been described, but most of the subjects have a solitary non-metastatic PHEO. Penetrance is unknown, probably low, and usually family history is not evocative.

4.7   *MAX* Gene

MAX mutations predispose to PHEO and abdominal PGL. Half of the patients develop bilateral disease. Penetrance is probably higher compared to *TMEM* mutations and about 40% of patients have a suggestive family history [12].

4.8   *FH* Gene

Mutations in *FH* genes were known to be associated with hereditary leiomyomatosis and papillary renal cell carcinoma; only recently mutations in this gene have been identified in PPGL. *FH* mutations predispose to multiple and/or malignant PPGL.

4.9   Other Genes

Application of exome whole-exome sequencing to PPGLs patients has led to identification of germline mutations in several genes; among them, some are of particular interest since they are involved in the hypoxia pathway (*EPAS1* and *EGLN1*), in mitochondrial metabolism (*MDH2*, *GOT2*, *SLC25A11*, *DLST*, *KIF1Bβ*), in the MAP kinase pathway (*MET* and *MERKT*) or in DNA methylation (*H3F3A*, *DNMT3A*) [6]. Currently, we have no clear knowledge about the real pathogenic role of these genes in the predisposition to PPGL, nor even elements to hypothesize penetrance and associated clinical phenotype.

4.10 Molecular Biology

Comprehensive genomic and transcriptome analysis has achieved the identification of three distinguishing molecular signatures in PPGL, each of them having germinal and/or somatic mutations in susceptibility genes leading to the activation of a specific oncogenic signaling pathway: the pseudo hypoxic, kinase, and Wnt signaling pathways. These three molecular clusters differ in phenotype and clinical behavior and, above all, they allow identification of molecular predictors of response to therapy, promising the application of personalized genetic-driven therapy in PPGL patients [6].

Cluster 1 is characterized by activation of the hypoxic pathway, and includes tumors with germinal/somatic mutations in the Krebs cycle-associated genes (*SDHx, FH, MDH2, GOT2, SLC25A11, DLST, IDH1*) and VHL/EPAS1-related genes (*PHD1/2, EGLN1, EPAS1, IRP1*). Mutations in these genes induce stabilization of HIF2α which in turn determines activation of angiogenesis, cell proliferation and migration; these tumors may have an aggressive phenotype, more than half of the patients with metastatic PPGL carry cluster 1 mutations. Cluster 1 tumors are mostly extra-adrenal and have a preference for noradrenaline secretion; *SDHx*-related PPGLs intensely express the somatostatin receptor 2 (SSTR2), then ^{68}Ga-DOTATATE PET/CT is considered the most sensitive functional imaging; on the contrary, *VHL/EPAS*-related tumors have lower expression of *SSTR2* and the most sensitive functional imaging is probably ^{18}F-DOPA PET/CT. New inhibitors of hypoxia-inducible factor 2-α represent an extraordinary opportunity for the therapy of metastatic forms.

Cluster 2 is characterized by activation of the tyrosine kinase-linked signaling pathway. It includes tumors with germline/somatic mutations in *NF1, RET, TMEM127, MAX, MET, MERTK, HRAS, FGFR1, B-RAF*. Mutations in these genes induce activation of phosphatidylinositol-3-kinase (PI3K)/AKT, mTORC and RAS/RAF/ERK signaling pathways leading to tumor proliferation, chromatin remodeling and angiogenesis. Cluster 2 tumors are mostly PHEO, have frequently an adrenergic phenotype and a less aggressive behavior compared to cluster 1. ^{18}F-DOPA/PET/TC is the preferable functional imaging due to the high uptake by the tumoral tissue compared to normal adrenal gland, achieving the detection of multiple lesions within the adrenal parenchyma. Tyrosine kinase inhibitors, PI3K/AKT/mTORC1 inhibitors and RAF/MEK/ERK inhibitors are all promising drugs for the systemic treatment in these tumors.

Cluster 3 is clearly not characterized from a molecular point of view, so it is much more difficult to identify unifying elements regarding clinical aspects. Until now only somatic driver mutations (*MAML3* fusion gene and *CSDE1* gene mutation) have been identified leading to overactivation of the Wnt signaling and β-catenin, which induces molecular events involved in carcinogenesis. Therapies targeting Wnt signaling are potentially suitable for tumors belonging to cluster 3.

References

1. Plouin PF, Amar L, Dekkers OM, et al. European Society of Endocrinology Clinical Practice Guideline for long-term follow-up of patients operated on for a phaeochromocytoma or a paraganglioma. Eur J Endocrinol. 2016;174(5):G1–G10.
2. Fassnacht M, Assie G, Baudin E, et al. Adrenocortical carcinomas and malignant phaeochromocytomas: ESMO-EURACAN Clinical Practice Guidelines for diagnosis, treatment and follow-up. Ann Oncol. 2020;31(11):1476–90. Erratum in: Ann Oncol. 2023;34(7):631.
3. Lenders JWM, Kerstens MN, Amar L, et al. Genetics, diagnosis, management and future directions of research of phaeochromocytoma and paraganglioma: a position statement and consensus of the Working Group on Endocrine Hypertension of the European Society of Hypertension. J Hypertens. 2020;38(8):1443–56.
4. Burnichon N, Rohmer V, Amar L, et al. The succinate dehydrogenase genetic testing in a large prospective series of patients with paragangliomas. J Clin Endocrinol Metab. 2009;94(8):2817–27.
5. Burnichon N, Mazzella JM, Drui D, et al. Risk assessment of maternally inherited SDHD paraganglioma and phaeochromocytoma. J Med Genet. 2017;54(2):125–33.
6. Nölting S, Bechmann N, Taieb D, et al. Personalized management of pheochromocytoma and paraganglioma. Endocr Rev. 2022;43(2):199–239. Erratum in: Endocr Rev. 2022;43(2):440; 437–9.
7. Gruber LM, Erickson D, Babovic-Vuksanovic D, et al. Pheochromocytoma and paraganglioma in patients with neurofibromatosis type 1. Clin Endocrinol (Oxf). 2017;86(1):141–9.
8. Kunst HP, Rutten MH, de Mönnink JP, et al. SDHAF2 (PGL2-SDH5) and hereditary head and neck paraganglioma. Clin Cancer Res. 2011;17(2):247–54.
9. Neumann HP, Bausch B, McWhinney SR, et al. Germ-line mutations in nonsyndromic pheochromocytoma. N Engl J Med. 2002;346(19):1459–66.
10. van der Tuin K, Mensenkamp AR, Tops CMJ, et al. Clinical aspects of SDHA-related pheochromocytoma and paraganglioma: a nationwide study. J Clin Endocrinol Metab. 2018;103(2):438–45. Erratum in: J Clin Endocrinol Metab. 2018;103(5):2077.
11. Tufton N, Ghelani R, Srirangalingam U, et al. SDHA mutated paragangliomas may be at high risk of metastasis. Endocr Relat Cancer. 2017;24(7):L43–9.
12. Buffet A, Burnichon N, Favier J, Gimenez-Roqueplo AP. An overview of 20 years of genetic studies in pheochromocytoma and paraganglioma. Best Pract Res Clin Endocrinol Metab. 2020;34(2):101416.

Imaging in Adrenocortical Carcinoma and Malignant Pheochromocytoma

5

Roberta Ambrosini, Francesco Bertagna, Francesco Dondi, Alessandro D'Amario, Teresa Falcone, and Luigi Grazioli

5.1   Introduction

The possibility of identifying primitive adrenal neoplasms at an early stage contrasts with the difficulty differentiating the nature of subclinical lesions, the so-called adrenal incidentalomas, i.e., lesions with a short axis ≥ 1 cm incidentally detected in non-oncological patients being examined for various reasons unrelated to the adrenal gland. These lesions, whose prevalence increases with age (up to 10% in patients aged 70 years), are mostly adenomas (70%) and, in 20% of cases, benign lesions of other nature (myelolipomas, benign pheochromocytomas, schwannomas, vascular lesions). Incidental malignant lesions account for approximately 10% and are most commonly represented by secondary lesions, adrenocortical carcinoma (ACC), pheochromocytoma (PHEO), and primary B lymphoma [1, 2]. It is therefore crucial to make a correct differential diagnosis between benign lesions, PHEOs with signs of malignancy, adrenocortical carcinomas and other primary/secondary lesions.

Approximately 70% of adrenal adenomas (AA) are represented by lipid-rich adenomas, which are characterized by high amounts of microscopic cytoplasmic fat, with attenuation values ≤ 10 HU (Hounsfield units) on unenhanced computed tomography (CT); moreover, they are often <4 cm in size and have a homogeneous structure. The very low attenuation value is considered very suggestive for AA (with

R. Ambrosini (✉) · A. D'Amario · T. Falcone · L. Grazioli
Diagnostic and Interventional Radiology Unit 1, ASST Spedali Civili di Brescia, Brescia, Italy
e-mail: robertambrosini@gmail.com; a.damario@unibs.it; t.falcone@unibs.it; lgrazioli@yahoo.com

F. Bertagna · F. Dondi
Nuclear Medicine, Department of Medical and Surgical Specialties, Radiological Sciences, and Public Health, University of Brescia at ASST Spedali Civili di Brescia, Brescia, Italy
e-mail: francesco.bertagna@unibs.it; f.dondi@outlook.com

© The Author(s) 2025
G. A. M. Tiberio (ed.), *Primary Adrenal Malignancies*, Updates in Surgery,
https://doi.org/10.1007/978-3-031-62301-1_5

sensitivity values of 55–71% and specificity of 98–100%) since attenuation $\leq$10 HU is rarely found in other adrenal neoplasms. Even the most recent guidelines of the European Society of Endocrinology (ESE) and the European Network for the Study of Adrenal Tumors (ENSAT) state that homogeneous lesions with HU <10 can exclude ACC with sufficient certainty. The remaining AA show attenuation $\geq$10 HU and are described as "lipid-poor". They are indistinguishable from other neoplasms on unenhanced CT, although they are less likely to be functioning [3].

If an adrenal nodule shows attenuation values >10 HU on unenhanced CT, a washout CT scan or magnetic resonance imaging (MRI) should be performed to confirm the possible adenomatous nature of the nodule. In nodules with attenuation values up to 20 HU, MRI with chemical-shift imaging (CSI) accurately assesses intracytoplasmic fat. At the same time, its use is considered inappropriate for lesions with 20–30 HU, since washout CT has shown 100% sensitivity in predicting lipid-poor adenomas, compared to 64% for CSI MRI. In addition, when a lesion has attenuation values $\geq$43 HU, an ^{18}F-FDG PET/CT may be more valuable, given the higher risk of malignancy [4]. Intralesional macroscopic fat (due to intratumoral adipocytes) has so far been considered a benign feature diagnostic of myelolipoma in accordance with the recommendations of the American College of Radiology [5]. However, because of the reported cases of adrenal neoplasms (including ACC) with macroscopic fat [6, 7], especially in lesions <4 cm, some authors suggest a cautious diagnostic approach, with a quantitative threshold of $\geq$50% macroscopic fat for this type of diagnosis.

In the case of adrenal incidentalomas, ^{18}F-FDG PET/CT has the advantage of distinguishing between benign and malignant tumors, although it does not provide robust information on the origin of the adrenal masses. With a sensitivity ranging between 86–100% and a specificity between 80–100% for the assessment of malignancy of adrenal lesions, depending on visual or semiquantitative assessment used, this imaging modality is helpful to rule out any diagnosis of ACC or metastatic disease [8]. Positive predictive value (PPV), negative predictive value (NPV) and accuracy have been reported as 81%, 100% and 95%, respectively [8–10]. Currently, the American College of Radiology recommends the use of ^{18}F-FDG PET/CT to define an indeterminate adrenal mass in patients with a history of cancer [5]. Radiolabeled metomidate has also a role for the assessment of adrenal incidentalomas, with sensitivity and specificity of 89% and 96%, respectively, in distinguishing adrenocortical from non-adrenocortical tumor masses [8–10].

5.2   Washout CT of Adrenal Lesions

The washout CT study is considered an essential test in the differential diagnosis between AA and primary or secondary adrenal neoplasms, with the possibility of calculating absolute and relative washout of contrast. The protocol includes the acquisition of unenhanced CT scans, followed by acquisitions in the portal-venous phase and late phase, respectively at 60–70 s and 15 min from contrast injection, with measurement of the lesion attenuation values and determination of the absolute (APW) and relative percentage washout (RPW) values using the following formulae:

$$APW = \frac{HU\ portal\ venous - HU\ delayed}{HU\ portal\ venous - HU\ unenhanced} \times 100$$

$$RPW = \frac{HU\ portal\ venous - HU\ delayed}{HU\ portal\ venous} \times 100$$

APW values ≥60% showed a sensitivity of 86–94% and a specificity of 92–96% in diagnosing AA, and RPW ≥40% showed a sensitivity of 96% and a specificity of 100% in diagnosing AA [11]. However, a recent multicenter study [12] analyzing a large and selected case series showed that the prevalence of malignant lesions in homogeneous nodules <4 cm did not differ significantly between the category with washout >60% and <60% (Fig. 5.1). This experience also showed that about 1/3 of PHEOs may show washout >60%; moreover, these enhancement features may also be observed in adrenal metastases from hepatocellular and renal cell carcinomas, which may show little fat component.

5.3   MRI of Adrenal Lesions

MRI of the adrenal glands relies on its intrinsic high contrast resolution and tissue characterization, particularly in detecting intracytoplasmic and macroscopic fat. The standard MRI protocol includes dual-echo T1-weighted (T1w) for CSI, T2-weighted (T2w) sequences, and T1w/T2w sequences with fat suppression.

In CSI, specific sequences can indicate a decrease in signal for the entire lesion or certain parts of it, which can reveal the amount of intracellular fat present. Microscopic fat content on MRI can be evaluated through two methods: qualitative visual assessment or quantitative measurements that involve placing regions of interest on in-phase and out-of-phase images. These measurements can be done with or without reference to the spleen signal, and allow for the calculation of signal intensity indices that reflect signal loss. These indices are represented by the adrenal-to-spleen CSI ratio (ASR) and the adrenal-signal-intensity index (ASII). In addition, T1w and T2w sequences with fat suppression are used in the MRI study protocol to assess macroscopic fat content (adipocyte aggregates or myelolipomatous portions). The addition of dynamic examination with paramagnetic contrast agents can further improve the diagnostic accuracy of MRI in diagnosing AA, with a sensitivity of 94% and specificity of 98% compared with 87% and 95% for imaging with CSI sequences alone. It is important to note that diffusion-weighted

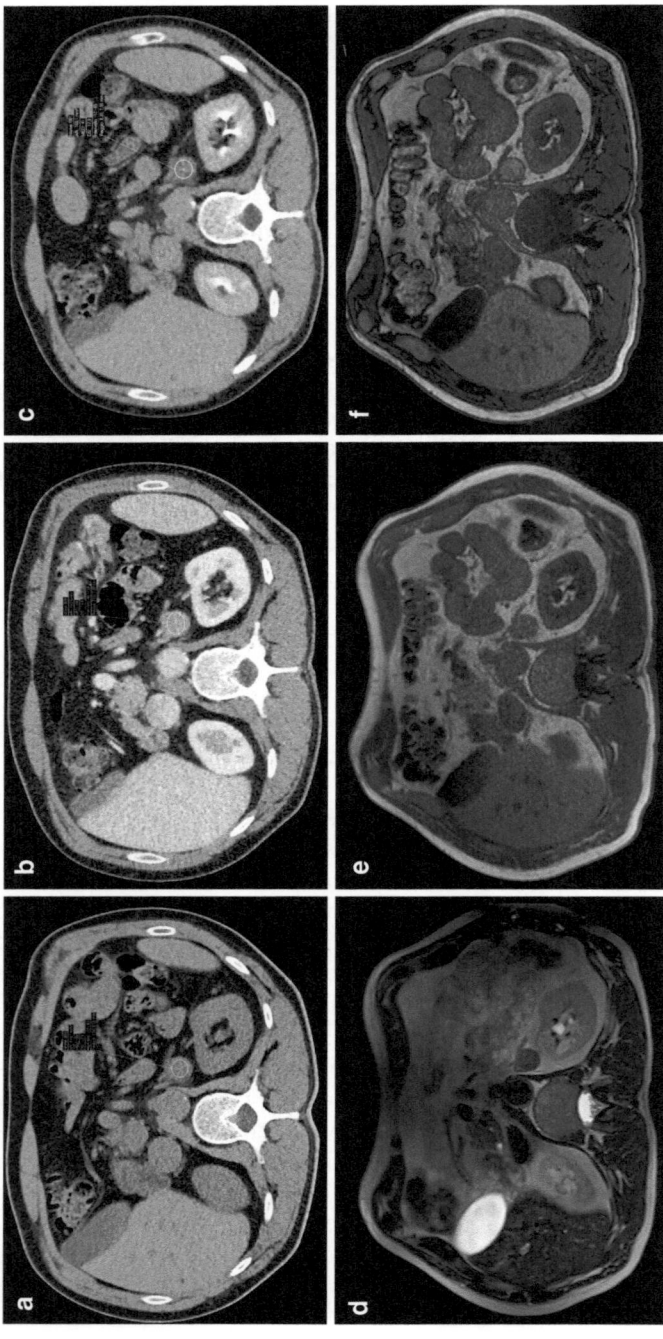

Fig. 5.1 Unenhanced CT in a 48-year-old patient with incidental detection of a 2 × 2 cm left adrenal lesion with attenuation values of 34 HU (**a**). The hormonal work-up was negative for hypersecretion and a washout CT (**b, c**) was performed to characterize the lesion, which showed an absolute and relative washout of 84% and 51%, respectively, suggesting a lipid-poor adrenal adenoma. After multidisciplinary discussion and in view of the high baseline attenuation of the lesion, a second imaging modality was suggested. On CSI MRI (**d, e, f**) the lesion showed no out-of-phase signal loss, with an adrenal/spleen CSI ratio of 1.06 (indeterminate if ≥0.71) and an ASII of 1.1% (indeterminate if ≤16.5%). PET-CT with FDG showed no pathological uptake by the lesion, and after a multidisciplinary case review with the patient, he preferred to wait for a follow-up unenhanced CT scan in 6 months rather than undergoing immediate surgery to rule out significant lesion growth (>20% maximum diameter in addition to at least 5 mm increase in maximum diameter), which has remained indeterminate to date

imaging (DWI) sequences alone cannot accurately determine the classification of adrenal lesions as benign or malignant. This is due to the fact that adenomas, which are typically benign, may still exhibit diffusion restriction on DWI [2, 3].

For incidental adrenal lesions <4 cm in size, without evidence of associated fat, without hormonal changes and not detected on PET scan, management is decided on a multidisciplinary basis, including the patient's preference. In this case, the lesion should be reassessed after 6–12 months with unenhanced imaging (CT or MRI) to assess any enlargement, in accordance with the ESE clinical practice guidelines [13] (Figs. 5.2 and 5.3).

5.4   Adrenocortical Carcinoma

ACC are often quite large at diagnosis (>6 cm in about 70%) and at advanced stages (18–26% stage III and 21–46% stage IV). Moreover, it is frequent to observe compression and dislocation caused by the mass on the adjacent organs and, occasionally, neoplastic thrombosis of the adrenal and renal veins, with possible extension to the inferior vena cava (IVC) or right atrium [8]. In about 15% of cases ACC can be found as a smaller, incidental lesion that requires further characterization. The few reports in the literature on early-stage ACC suggest that, regardless of size, the following radiological features may serve as criteria for an early diagnosis: lesions with high attenuation values (≥30 HU), calcifications (30% of cases), irregular shape, inhomogeneous structure and poorly defined contours, although there are no imaging-specific features [2].

According to the ENSAT guidelines, contrast-enhanced CT (CECT) of the chest, abdomen and pelvis is the imaging modality of choice for initial staging and follow-up during and after treatment. At the same time, MRI and PET/CT may provide

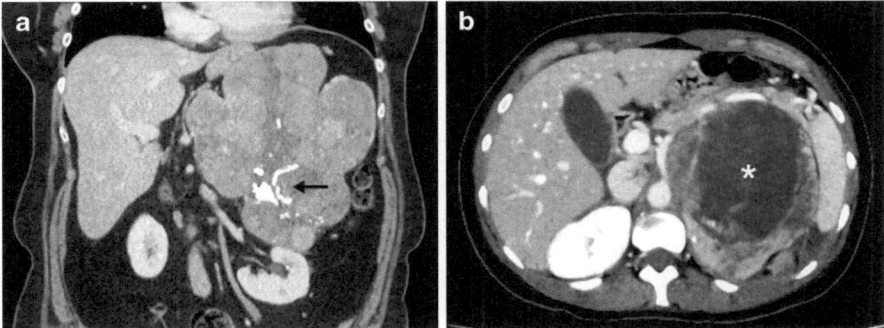

Fig. 5.2 (**a**) Coronal venous phase CT in a 38-year-old male shows a heterogeneous ACC in the left adrenal loggia, displacing the left kidney inferiorly. Amorphous calcifications are also visible (*black arrow*). (**b**) 28-year-old woman who underwent a CT scan due to abdominal pain. The axial CT venous phase shows an inhomogeneous mass in the left adrenal fossa with peripheral hypervascularity due to the presence of solid tissue and a central hypodense area secondary to necrosis (*asterisk*). The lesion displaces the spleen and the pancreatic tail anteriorly

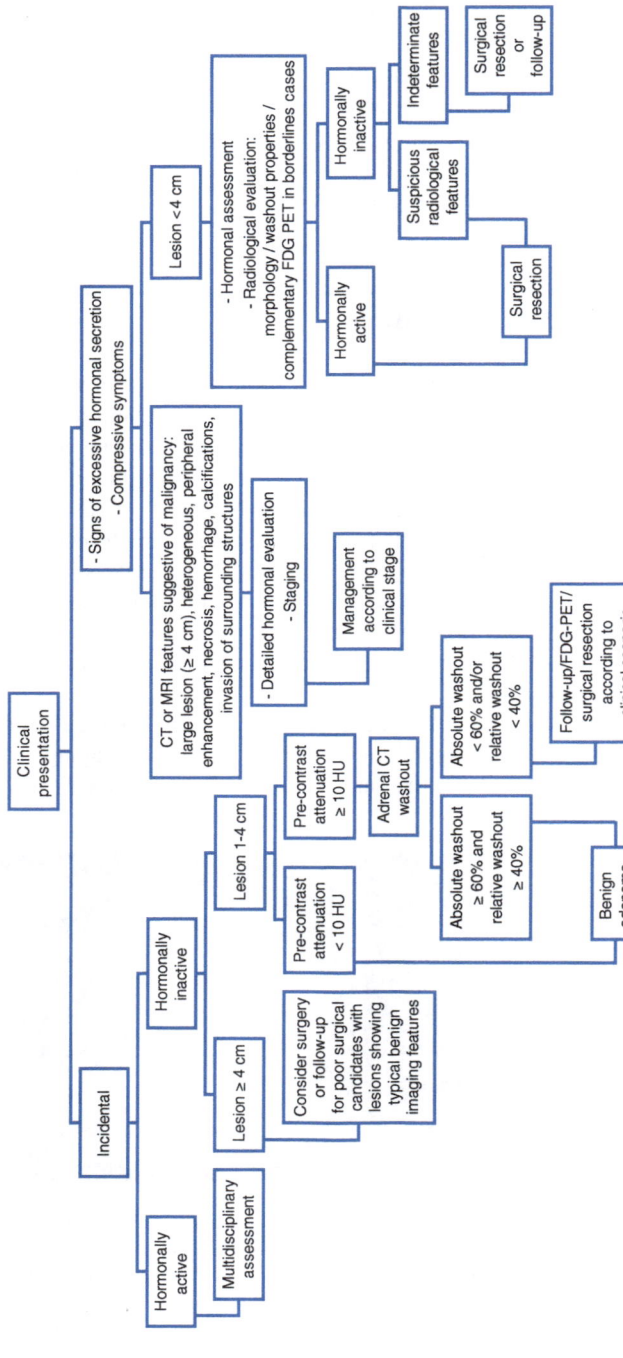

Fig. 5.3 Summary of diagnostic work-up for the management of incidental adrenal lesions or lesions with evidence of hormonal hypersecretion and/or pressure symptoms, according to European Society of Endocrinology and European Network for the Study of Adrenal Tumors guidelines

additional information in selected cases [14]. CECT is performed not only to assess adrenal lesion vascularity, but also (especially in large tumors) to define local relationships, infiltration of adjacent organs, endovascular extension, peritoneal spread, lymphadenopathies or distant metastases.

ACC typically appears inhomogeneous on unenhanced CT, with hypoattenuating areas of necrosis or cystic degeneration and slightly hyperattenuating components from hemorrhagic events or calcifications. In CECT the tumor is equally inhomogeneous, with thicker peripheral solid tissue ("rim enhancement") and central necrotic portions (Fig. 5.4). Calculation of APW and RPW in large lesions is not relevant, but in a smaller lesion with a homogeneous structure these would be <60% and <40%, respectively. Similarly, on MRI, the signal is typically inhomogeneous, iso/hypointense to the liver parenchyma on T1w sequences, except in hemorrhagic areas where the signal is slightly hyperintense. T2w sequences are the best ones to emphasize cystic degeneration or necrotic areas. Prominent and inhomogeneous enhancement with slow washout follows the injection of gadolinium-chelate contrast agents.

Beyond the depiction of pathologic lymph nodes and metastases, preoperative imaging is essential to define the presurgical assessment of the patient. Indeed, it is essential to evaluate the infiltration of renal and hepatic parenchyma and the possible presence of neoplastic thrombosis of the adrenal and renal vessels and involvement of the IVC. Both CT and MRI have a high sensitivity and specificity in the assessment of these diagnostic details, even though MRI has been shown to be superior to CT in detecting IVC invasion and assessing its extension (with respect to the hepatic veins' confluence and right atrium) due to its intrinsic multiplanarity and contrast resolution. Hepatic parenchymal infiltration by the mass must be suspected in the presence of neoplastic involvement of periadrenal fat, disappearance or reduction <1 mm of the fat line between the liver and the neoplasm, compression and mass effect on the IVC or right hepatic lobe, disruption of the adrenal capsule, enhancement of the periadrenal liver parenchyma, focal bulging of the ACC towards the liver or inclusion of the mass by the liver parenchyma >180° [15].

Metastatic diffusion is often already present at the time of diagnosis, resulting in enlarged para-aortic lymph nodes (25–46%), pulmonary lesions (45–97%) and liver metastases (48–96%). Hepatic metastases (especially when small) tend to be hypervascular with subsequent washout, and are best appreciated in the arterial phase following contrast injection, which is why this phase is considered necessary for follow-up imaging. The use of contrast-enhanced MRI using hepatospecific gadolinium-chelates (such as Gd-BOPTA or Gd-EOB-DTPA) may be useful for a more accurate staging of secondary hepatic involvement, allowing an accurate differential diagnosis between metastases and benign hypervascular lesions (in particular, focal nodular hyperplasia-like regenerative nodules associated with long-term chemotherapy or other benign lesions).

The assessment of response to therapy in locally advanced or metastatic disease still relies on the RECIST 1.1 criteria, which are based on dimensional changes in "target" lesions. However, tumors (and metastases) do not always have standard volumes, and different areas of the same tumor may respond differently to therapy, with changes in size resulting from therapy-induced necrosis. For this

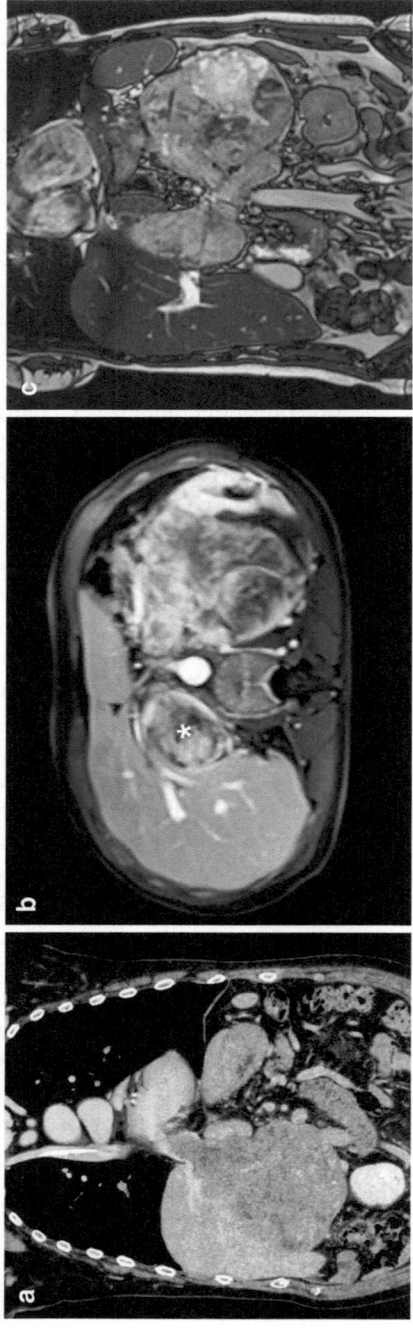

Fig. 5.4 (**a**) Coronal venous phase CT scan in a 72-year-old man shows a large inhomogeneous mass (ACC) invading the inferior vena cava and inseparable from the liver. Axial 3D GRE fat-suppressed T1w after contrast administration (**b**) and coronal T2w (**c**) MRI sequence in a 52-year-old woman show enlarged left renal vein and inferior vena cava (*asterisk*) due to the presence of heterogeneous intraluminal neoplastic tissue originating from the voluminous mass located in the left adrenal fossa (ACC). Note the intrinsically better resolution of contrast-enhanced MRI compared to CT

reason, in recent years it has been proposed to consider not only the size criterion, but also changes in vascularity, assessed as variations in enhancement in HU (based on the Choi criteria), as well as volumetric changes of target lesions. In our experience, we have also shown that concordance of all three criteria is associated with better outcomes and overall survival [16]. In ACC, ^{18}F-FDG PET/CT has the ability to evaluate the overall extent of disease and the presence of metastases, thus guiding the management of the patient, even though false negative results have been reported in 11% of cases [17]. Reported sensitivity, specificity, PPV, NPV and accuracy values are 100%, 40–95%, 71%, 100% and 76%, respectively, for the diagnosis of ACC [18, 19]. Moreover ^{18}F-FDG PET/CT can be helpful to predict the metabolic response of ACC, and some insights on its prognostic role have emerged. In this setting, patients with higher tracer uptake seem to be characterized by a poorer prognosis, although the reported findings are conflicting [17]. Metomidate (METO) is the methyl ester of etomidate inhibitors of the CYP11β enzymes (11β-hydroxylase and aldosterone synthase) and can be labeled with different isotopes (^{123}I, [124]I or ^{11}C) and used therefore to image the adrenal cortex with scintigraphy or PET/CT. In this setting, ^{11}C-METO has a sensitivity of 72% for the assessment of ACC [19].

5.5 Malignant Pheochromocytoma

The main imaging criteria suggestive of malignant PHEOs include large size, evidence of infiltration of adjacent organs, and lymphadenopathy. All PHEOs may have malignant potential, which is not detectable on imaging, especially in lesions ≤4 cm that do not show local invasion or metastasis at diagnosis. For this purpose, several scoring systems have been proposed to predict the malignant potential of the lesion. The two most widely accepted are PASS (Pheochromocytoma of the Adrenal Gland Scaled Score) and GAPP (Grading System for Adrenal Pheochromocytoma and Paraganglioma). These grading systems consider as possible features of malignancy: attenuation >10 HU, even in heterogeneous lesions with cystic (approximately 7% of adrenal cysts are PHEOs) or necrotic/hemorrhagic portions and calcifications (in 20% of cases) [20]. It is important to note that on washout CT, at least 35% of PHEOs may present with APW and RPW similar to those of lipid-poor AA. The abundant and disordered blood supply determines the pronounced enhancement in the arterial and venous phase, which is greater than that observed in lipid-poor AA. A recent paper published by the Society of Abdominal Radiology [21] has proposed the use of a venous-phase attenuation threshold of 130 HU, which has a high specificity for the diagnosis of PHEO (100%), although with sensitivity values of only 38%.

On MRI, PHEOs may show very high signal on T2w sequences (known as the "light bulb sign"), which is present in only 50% of PHEOs; in other cases they show generally higher T2w signal compared to the contralateral adrenal gland and lipid-poor adenomas. These lesions do not have intracytoplasmic fat and therefore appear slightly hypointense on T1w sequences and do not show a signal loss on CSI (Fig. 5.5). DWI has a role only in the detection of PHEOs <1 cm [20].

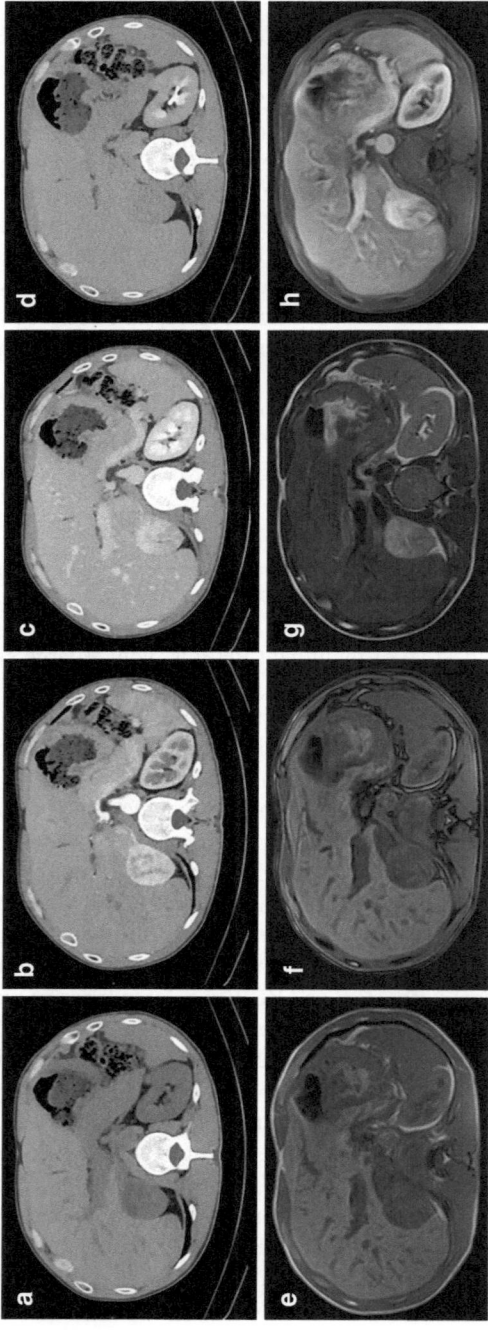

Fig. 5.5 36-year-old man with an incidental adrenal mass discovered on ultrasound at an outside center. Unenhanced axial CT (**a**) shows a well-circumscribed oval mass with a maximum size of 5.3 cm involving the right adrenal gland with pre-contrast attenuation of 34 HU. Axial contrast-enhanced arterial phase CT (**b**) shows early enhancement with high lesion attenuation (122 HU) in the portal venous phase (**c**). On 15-min delayed CT scans (**d**), the lesion shows an attenuation value of 65 HU, resulting in absolute and relative washout values of 65% and 47%, respectively. In view of these washout values and after multidisciplinary case discussion, MRI was performed to better define the lesion characteristics and to try to rule out a lipid-poor adenoma in a patient with elevated catecholamines on hormonal work-up. Axial out-of-phase and in-phase T1w CSI GRE images (**e, f**) show that the mass has no decrease in signal intensity, with high signal on axial T2w images, also known as the "light bulb sign" (**g**). Also on contrast-enhanced MRI, the lesions showed bright and early contrast enhancement (**h**). The mass was surgically resected and pathologically proven to be a pheochromocytoma

5.6   Nuclear Medicine and Pheochromocytoma

Molecular imaging can be useful in PHEO to assess the presence of disease when borderline metanephrine levels and indeterminate adrenal masses are present, for the assessment of locoregional and distant extension of disease, for the assessment of its aggressiveness, the detection of therapeutic molecular targets and to evaluate the response to therapy [22]. The European Association of Nuclear Medicine (EANM) guidelines recommend to perform nuclear imaging for PHEO in the case of large tumors (>5 cm), succinate dehydrogenase subunit B (SDHB) mutated status, noradrenergic biochemical phenotype, and/or high methoxytyramine levels [22, 23]. Interestingly, most published studies considered nuclear medicine modalities for the evaluation of both PHEO and paraganglioma (PGL)—together referred to as PPGL. Moreover, pregnancy and breastfeeding are general contraindications for molecular imaging procedures. ^{111}In-DTPA-pentetreotide specifically binds to somatostatin receptors (SSTR) and can be therefore used to visualize PHEO with scintigraphy or single photon-emission CT (SPECT), although with low overall sensitivity (30%). Moreover, considering the high radiation exposure, cost and long waiting periods for imaging, its use has fallen out of favor [24].

Metaiodobenzylguanidine (MIBG) is an analog of norepinephrine taken up by norepinephrine transporters and then stored in neurosecretory vesicles of sympathetic presynaptic neurons [23]. ^{123}I-MIBG scintigraphy or SPECT/CT is useful for the assessment of PHEO in terms of disease burden, with sensitivities and specificities ranging between 83–100% and 70–100%, respectively [24]. ^{131}I-MIBG can also be used to image PHEO, but it has some dosimetric and resolution issues that discourage its use. In this setting, ^{123}I-MIBG scintigraphy is therefore useful for the selection of potential candidates for ^{131}I-MIBG radiometabolic therapy [22]. Thyroid blockade with administration of 130 mg of potassium iodide should be performed 1 h before tracer injection in the case of ^{123}I-MIBG or 24 h before and continued daily for at least 5 days for ^{131}I-MIBG. Many drugs modify the uptake of MIBG and have to be suspended before scintigraphy. Rare adverse events such as tachycardia, pallor or vomiting can be experienced by the patients during administration of the tracer and can be prevented by slow injection [23]. ^{123}I-MIBG scintigraphy should not be used in patients with SDHB mutation, hereditary, extra adrenal and metastatic PPGL, which exhibit only a low rate of positivity [22]. In contrast, this imaging modality is useful in patients with negative genetic screens, those with bilateral adrenal lesions, suspicious conventional imaging scans and subjects with biochemical suspicion of PHEO [25]. The general sensitivity of SPECT/CT is hampered by a low spatial resolution and therefore ^{124}I-MIBG PET/CT is emerging as a new imaging modality for the assessment of PHEO, although with reported similar accuracy to $^{123/131}$I-MIBG scintigraphy [25].

^{68}Ga-DOTANOC, ^{68}Ga-DOTATOC and ^{68}Ga-DOTATATE are labeled SSTR analogues (SSA) used in PET/CT that can contribute to PPGL detection, in particular when metastatic or extra-adrenal diseases are present, with an overall sensitivity ranging between 80–100% and a detection rate between 93–98% [22]. Notably, DOTATATE binds most avidly to SSTR2, the most common somatostatin receptor expressed in PPGLs, and is therefore the most widely used in particular for cluster 1A PHEO (especially SDHx), and metastatic and pediatric PPGL [23]. The sensitivity of ^{68}Ga-SSA PET/CT has been reported to be 94% for pediatric SDHx-related diseases, 99% for metastatic SDHB-related and SDHD-related PPGLs, and 100% for SDHA-related neoplasm. Moreover, it can also be used to determine whether a patient is likely to benefit from peptide receptor radionuclide therapy and, in the case of restaging, it can change the management in most of the patients [22].

^{18}F-FDG has high sensitivity in the detection of metastatic PPGL, in particular for SDHB patients, with reported values ranging between 77–100% [25]. However, PHEO usually has increased but variable tracer uptake and reported specificity, PPV and NPV for PET/CT near 96% [22, 23]. Interestingly, MEN-2-associated PHEOs are ^{18}F-FDG-avid in only 40% of the cases, and cluster 1 patients have higher uptake compared to cluster 2 subjects. As a consequence, ^{18}F-FDG PET/CT should be considered in the preoperative workup of PPGL, in particular in SDHB metastatic neoplasms [22, 23].

[18]F-L-dihydroxyphenylalanine (^{18}F-DOPA) is taken up by L-type amino-acid transporter (LAT) and is stored in neurosecretory granules of catecholamine-producing cells. Therefore, high sensitivity and specificity in the detection of non-metastatic PHEO have been reported, with sensitivity of 94% in patients of known genetic background and 100% in apparently sporadic non-metastatic PHEO [23, 25]. In the case of metastatic disease, ^{18}F-DOPA PET/CT was found to perform better for SDHB-negative PPGLs (93% sensitivity) than for SDHB-positive cases (20% sensitivity), where ^{18}F-FDG uptake may be higher. Furthermore, high sensitivity for the detection of VHL-, EPAS1 (HIF2A)-, and FH-associated PPGLs were reported [22]. Interestingly, some authors suggest the administration of 200 mg of carbidopa 1 h before the examination to block decarboxylation of DOPA to dopamine and improve the uptake in target tissues [23].

5.7   Comparison Between Nuclear Medicine Modalities

Generally speaking, PET/CT technology has been shown to be superior to SPECT/CT and scintigraphy, with higher spatial resolution, greater sensitivity and fewer indeterminate or equivocal findings. Furthermore, in PHEO patients ^{123}I-MIBG has significantly outperformed ^{111}In-pentetreotide in the detection of disease. Due to its nonspecific and variable accumulation, ^{18}F-FDG is not considered the tracer of choice for PPGLs imaging, although with superior sensitivity compared to ^{123}I-MIBG imaging in SDHx-related and metastatic tumors [22].

According to the latest EANM guidelines, in the case of sporadic PHEO, the first-choice imaging modality should be ^{18}F-DOPA PET/CT or ^{123}I-MIBG SPECT/CT, the second one should be ^{68}Ga-SSA PET/CT and the third should be ^{18}F-FDG PET/CT. As for inherited PHEO (NF1, RET, VHL and MAX) with the exception of SDHx, the first choice should be ^{18}F-DOPA PET/CT, the second ^{123}I-MIBG SPECT/CT or ^{68}Ga-SSA PET/CT while the third should be ^{18}F-FDG PET/CT. In the case of extra-adrenal sympathetic and/or multifocal and/or metastatic SDHx mutation, the first choice should be ^{68}Ga-SSA PET/CT, the second choice ^{18}F-FDG PET/CT or ^{18}F-DOPA PET/CT and the third one should be ^{18}F-FDG PET/CT and ^{123}I-MIBG SPECT/CT or ^{18}F-FDG PET/CT and ^{111}In-DTPA-pentetreotide SPECT/CT. Moreover, all the third choices should be considered only if ^{18}F-DOPA and ^{68}Ga-SSA PET/CT are not available. Interestingly, in cluster 3 subjects, the most sensitive functional imaging modality is unknown (Fig. 5.6) [22, 23, 25].

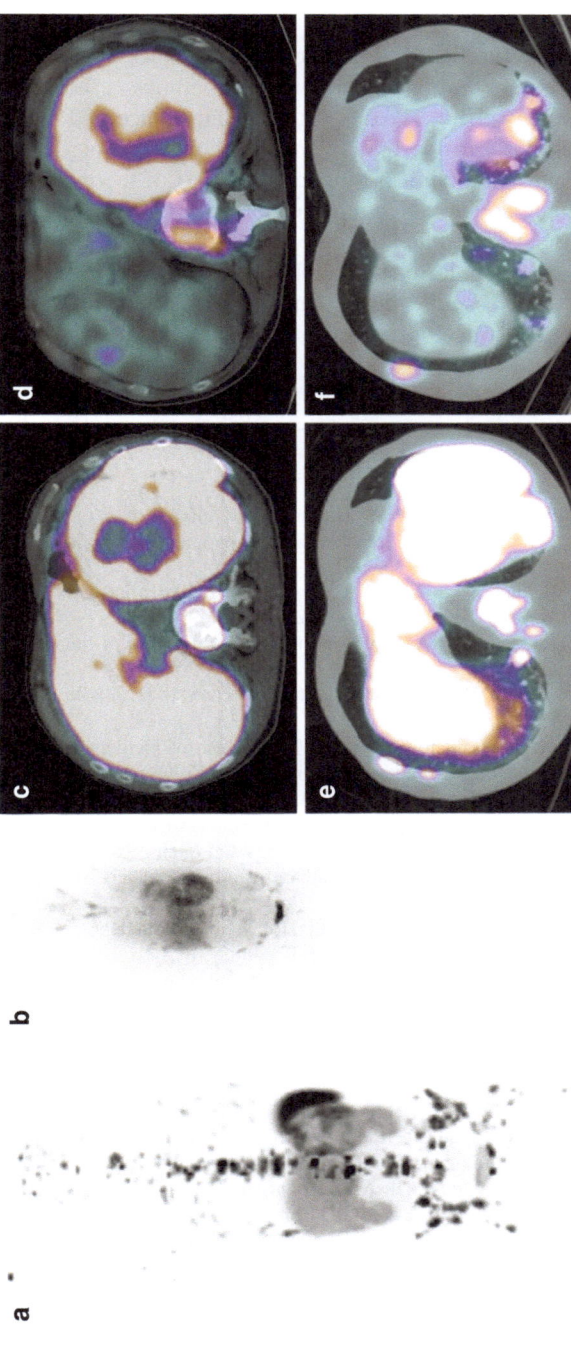

Fig. 5.6 Maximum intensity projection (**a**) and axial fused PET/CT images (**c**, **e**) of a ^{68}Ga-DOTATOC scan performed for staging in a patient with pheochromocytoma, showing intense uptake by the primary lesion and multiple skeletal and pulmonary metastases. The same findings were confirmed on anterior view (**b**) and axial fused SPECT/CT images (**d**, **f**) of a 123I-MIBG scan of the same patients, but without demonstration of pulmonary uptake

5.8 Conclusions

The radiological and nuclear medicine imaging of adrenal lesions still represents a diagnostic challenge, where the role of the different methods is seen as a "link" in the wider multidisciplinary chain. The integration of unenhanced CT, washout study and MRI features, similarly to the findings of molecular imaging methods, must therefore be considered in the clinical-functional context of the individual patient. Many radiomics studies are investigating the potential of more in-depth texture analysis at CT and MRI, which could increase accuracy in the differential diagnosis between benign and malignant lesions. In addition, the application and standardization of radiogenomic analyses may, in the near future, allow more precise and personalized management of oncological and surgical treatments for better patient outcomes.

References

1. Song JH, Chaudhry FS, Mayo-Smith WW. The incidental adrenal mass on CT: prevalence of adrenal disease in 1,049 consecutive adrenal masses in patients with no known malignancy. AJR Am J Roentgenol. 2008;190(5):1163–8.
2. Barat M, Cottereau AS, Gaujoux S, et al. Adrenal mass characterization in the era of quantitative imaging: state of the art. Cancers (Basel). 2022;14(3):569.
3. Bracci B, De Santis D, Del Gaudio A, et al. Adrenal lesions: a review of imaging. Diagnostics (Basel). 2022;12(9):2171.
4. Nandra G, Duxbury O, Patel P, et al. Technical and interpretive pitfalls in adrenal imaging. Radiographics. 2020;40(4):1041–60.
5. Mayo-Smith WW, Song JH, Boland GL, et al. Management of incidental adrenal masses: a white paper of the ACR Incidental Findings Committee. J Am Coll Radiol. 2017;14(8):1038–44.
6. Egbert N, Elsayes KM, Azar S, Caoili EM. Computed tomography of adrenocortical carcinoma containing macroscopic fat. Cancer Imaging. 2010;10(1):198–200.
7. Zhang HM, Perrier ND, Grubbs EG, et al. CT features and quantification of the characteristics of adrenocortical carcinomas on unenhanced and contrast-enhanced studies. Clin Radiol. 2012;67(1):38–46.
8. Viëtor CL, Creemers SG, van Kemenade FJ, et al. How to differentiate benign from malignant adrenocortical tumors? Cancers (Basel). 2021;13(17):4383.
9. Mendichovszky IA, Powlson AS, Manavaki R, et al. Targeted molecular imaging in adrenal disease—an emerging role for metomidate PET-CT. Diagnostics (Basel). 2016;6(4):42.
10. Shariq OA, McKenzie TJ. Adrenocortical carcinoma: current state of the art, ongoing controversies, and future directions in diagnosis and treatment. Ther Adv Chronic Dis. 2021;12:20406223211033103.
11. Caoili EM, Korobkin M, Francis IR, et al. Adrenal masses: characterization with combined unenhanced and delayed enhanced CT. Radiology. 2002;222(3):629–33.
12. Corwin MT, Navarro SM, Malik DG, et al. Differences in growth rate on CT of adrenal adenomas and malignant adrenal nodules. AJR Am J Roentgenol. 2019;213(3):632–6.
13. Fassnacht M, Tsagarakis S, Terzolo M, et al. European Society of Endocrinology clinical practice guidelines on the management of adrenal incidentalomas, in collaboration with the European Network for the Study of Adrenal Tumors. Eur J Endocrinol. 2023;189(1):G1–G42.
14. Fassnacht M, Dekkers OM, Else T, et al. European Society of Endocrinology Clinical Practice Guidelines on the management of adrenocortical carcinoma in adults, in collaboration with the European Network for the Study of Adrenal Tumors. Eur J Endocrinol. 2018;179(4):G1–G46.

15. Kedra A, Dohan A, Gaujoux S, et al. Preoperative detection of liver involvement by right-sided adrenocortical carcinoma using CT and MRI. Cancers (Basel). 2021;13(7):1603.
16. Ambrosini R, Balli MC, Laganà M, et al. Adrenocortical carcinoma and CT assessment of therapy response: the value of combining multiple criteria. Cancers (Basel). 2020;12(6):1395.
17. Satoh K, Patel D, Dieckmann W, et al. Whole body metabolic tumor volume and total lesion glycolysis predict survival in patients with adrenocortical carcinoma. Ann Surg Oncol. 2015;22(Suppl 3):S714–20.
18. Krishnaraju VS, Kumar R, Subramanian K, et al. Fluoro-2-deoxyglucose-positron emission tomography/computed tomography in the diagnosis and management of adrenocortical carcinoma: a 10-year experience from a tertiary care institute. Indian J Nucl Med. 2022;37(3):227–35.
19. Ahmed AA, Thomas AJ, Ganeshan DM, et al. Adrenal cortical carcinoma: pathology, genomics, prognosis, imaging features, and mimics with impact on management. Abdom Radiol (NY). 2020;45(4):945–63.
20. Gruber LM, Strajina V, Bancos I, et al. Not all adrenal incidentalomas require biochemical testing to exclude pheochromocytoma: Mayo clinic experience and a meta-analysis. Gland Surg. 2020;9(2):362–71.
21. Glazer DI, Mayo-Smith WW, Remer EM, et al. Lexicon for adrenal terms at CT and MRI: a consensus of the Society of Abdominal Radiology adrenal neoplasm disease-focused panel. Abdom Radiol (NY). 2023;48(3):952–75.
22. Taïeb D, Hicks RJ, Hindié E, et al. European Association of Nuclear Medicine Practice Guideline/Society of Nuclear Medicine and Molecular Imaging Procedure Standard 2019 for radionuclide imaging of phaeochromocytoma and paraganglioma. Eur J Nucl Med Mol Imaging. 2019;46(10):2112–37.
23. Ryder SJ, Love AJ, Duncan EL, Pattison DA. PET detectives: molecular imaging for phaeochromocytomas and paragangliomas in the genomics era. Clin Endocrinol (Oxf). 2021;95(1):13–28.
24. Anyfanti P, Mastrogiannis K, Lazaridis A, et al. Clinical presentation and diagnostic evaluation of pheochromocytoma: case series and literature review. Clin Exp Hypertens. 2023;45(1):2132012.
25. Sbardella E, Grossman AB. Pheochromocytoma: an approach to diagnosis. Best Pract Res Clin Endocrinol Metab. 2020;34(2):101346.

Management of Endocrine Syndromes Associated with Adrenocortical Carcinoma

6

Chiara Borin, Soraya Puglisi, Anna Calabrese, Paola Perotti, and Massimo Terzolo

6.1 How Often Will I Find an Adrenocortical Carcinoma with Associated Endocrine Syndromes in My Practice?

Interestingly, 50–60% of adrenocortical carcinoma (ACC) patients present with steroid hypersecretion at diagnosis, which may result in overt endocrine syndromes, which cause additional challenges to the management [1, 2]. In a large and well-characterized series of patients with ACC from 12 expert centers in Italy, steroid hypersecretion was reported in more than half of cases and hypercortisolism was the most common hormonal excess in both sexes [3]. In women with ACC, the second most common isolated hypersecretion was androgen excess, while ACC secreting aldosterone or estrogens in men were uncommon [3]. Secretion of multiple steroids (usually cortisol plus androgens, or cortisol and other steroids) is a frequent condition, which can be considered a hallmark of ACC [1].

6.2 What Is the Clinical Presentation of Adrenocortical Carcinoma with Associated Endocrine Syndromes?

Hypercortisolism can occur with a large spectrum of clinical manifestations, from subclinical to overt Cushing's syndrome (CS), depending on the severity and the duration of glucocorticoid excess.

In patients with overt CS, physical examination can reveal central obesity with muscle hypotrophy of the upper and lower limbs, round face (the so-called "moon-like face"), a hump on the back of the neck ("buffalo hump"), thin skin with purple

C. Borin · S. Puglisi (✉) · A. Calabrese · P. Perotti · M. Terzolo
Internal Medicine, Department of Clinical and Biological Sciences, University of Turin, San Luigi Gonzaga University Hospital, Orbassano (Turin), Italy
e-mail: chiara.borin@unito.it; soraya.puglisi@unito.it; anna.calabrese678@gmail.com; paola.perotti@unito.it; massimo.terzolo@unito.it

© The Author(s) 2025
G. A. M. Tiberio (ed.), *Primary Adrenal Malignancies*, Updates in Surgery,
https://doi.org/10.1007/978-3-031-62301-1_6

striae and widespread ecchymoses. When faced with this full-blown picture, often associated with severe metabolic alterations (hypokalemia, hyperglycemia, metabolic alkalosis), the diagnosis of CS is fairly obvious [4]. However, other patients could present with subtle cortisol secretion, without overt stigmata. Even patients with subclinical CS can show detrimental consequences such as hypertension, insulin resistance with glucose intolerance or type 2 diabetes mellitus, hirsutism and menstrual irregularities (in women), neuropsychological symptoms (including depression, insomnia, irritability and memory loss), proximal myopathy and osteoporosis with an increased risk of bone fractures. A high frequency of thromboembolic events has also been reported. This is due to a hypercoagulability status that contributes, in addition to the abovementioned cardiometabolic comorbidities, to produce an increased risk of cardiovascular events. Finally, the chronic exposure to hypercortisolism causes immunodepression with a consequent increased risk of bacterial, viral and fungal infections [5]. Therefore, patients with cortisol-secreting ACC are complex and fragile, and they need a multidisciplinary approach for optimal management.

Androgen excess in women can cause hirsutism, acne and signs of virilization, such as deepening of voice and temporal balding, which only occur in females with severely high levels of serum androgens. Estrogen excess can cause testicular atrophy and gynecomastia in men through the induction of hypogonadism. The excess of mineralocorticoids is characterized by severe hypokalemia and hypertension [6].

6.3 How to Diagnose the Endocrine Syndrome Associated with Adrenocortical Carcinoma?

As recommended by the European Society of Endocrinology (ESE) and European Network for the Study of Adrenal Tumors (ENSAT) guidelines, in all patients with a suspected ACC a careful physical examination aiming to identify suggestive clinical features of steroid excess and a hormonal work-up should be performed before surgery [2]. The key points to assess the hormonal activity of adrenal masses are reported in the chapter on adrenal incidentaloma (see Ch. 8).

In the case of a suspected ACC, the hormonal work-up should also include assessment of serum dehydroepiandrosterone sulfate (DHEAS), 17-OH progesterone, androstenedione, 17β-estradiol (only in men and postmenopausal women), and testosterone (only in women).

The identification of hormonal hypersecretions is a key step in the management of patients with ACC, particularly in the case of hypercortisolism. Severe cortisol excess can cause life-threating complications (hypokalemia, infections, thromboembolic events) that should be recognized and treated as soon as possible, requiring a medical treatment which could impact the chances of patient survival. Moreover, patients with cortisol-secreting ACC who underwent surgery could develop postoperative adrenal insufficiency. Finally, monitoring of the steroid profile during follow-up after radical resection may be useful to detect ACC recurrence.

6.4   Does Hypercortisolism Affect Patient Outcome?

Cortisol excess is a well-known negative prognostic factor in patients with ACC, in addition to tumor stage, resection status and Ki67 index [7–10].

In recent decades, retrospective studies [11–14] and a meta-analysis [15] showed that hypercortisolism was a strong independent factor associated with shorter overall and recurrence-free survival. Conversely, no association between androgen excess and survival has been found (relative risk 0.82, 95% CI 0.60–1.12) [15].

Interestingly, a recent multicenter study demonstrated that at multivariate analysis hypercortisolism was associated with a higher risk of recurrence (hazard ratio [HR] 2.17; 95% CI 1.50–3.12; $p < 0.001$] in patients with localized ACC, and with a high risk of mortality both in patients with localized (HR 2.15; 95% CI 1.34–3.46; $p = 0.002$) and metastatic (HR 2.05; 95% CI 1.06–3.98; $p < 0.05$) disease at diagnosis [3].

However, the exact mechanisms responsible for this detrimental impact of hypercortisolism on prognosis are largely unknown. Many factors can be considered. First, severe hypercortisolism causes several potentially life-threating complications, as previously discussed. Moreover, cortisol excess has an immunosuppressive effect which may favor tumor progression. Another reasonable hypothesis is that functional tumors are more aggressive. This concept is supported by retrospective studies reporting higher Ki67 index in this subset of tumors compared to non-secreting ACC [3, 9] and by genomic studies demonstrating a transcriptome signature in cortisol-secreting tumors linked to a more aggressive behavior [16].

6.5   Which Treatment for Endocrine Syndromes?

When addressing hypercortisolism in patients with ACC, it is important to implement two parallel strategies: on one hand, a prompt control of cortisol excess and, on the other hand, treatment and prevention of the complications of CS.

Adrenalectomy is the mainstay treatment of ACC and aims to achieve complete resection in patients with localized disease or debulking in metastatic patients. Nonetheless, it is sometimes necessary to control hypercortisolism with medical treatment before surgery to reduce the risk of perioperative and postoperative complications. Presurgical control of cortisol excess is mandatory when the operative risk is presumed to be elevated because of poor general condition and/or severe hypercortisolism. A nationwide analysis including 276 patients with functional ACC (86% cortisol, 16% aldosterone, and 4% androgen excess) showed a significantly increased risk of wound issues, adrenocortical insufficiency and acute kidney injury in the postoperative period and a longer hospital stay compared with 1923 patients with non-functional ACC [17]. Similarly, a retrospective study, analyzing perioperative complications and mortality within 30 days from adrenalectomy, reported longer hospital stay and increased frequency and severity of complications in patients with cortisol-secreting compared with non-cortisol-secreting ACC [14].

The other cornerstone of treatment in ACC is mitotane, a cytotoxic drug that is also able to inhibit several enzymes of adrenal steroidogenesis. Therefore, mitotane reduces cortisol secretion, but it has a slow onset of action and it could require months before reaching full efficacy which is linked to therapeutic plasma concentrations [18]. To promptly control severe hypercortisolism, steroidogenesis inhibitors with rapid onset of action can be used. Ketoconazole, metyrapone and osilodrostat are administered orally, while etomidate is administered intravenously.

Moreover, in patients with severe hypercortisolism, a combined therapy could be useful. The combination of metyrapone and ketoconazole with mitotane was reported to rapidly control severe hypercortisolism in eight patients with ACC, leading to a dramatic reduction of cortisol secretion during the first week of treatment and a rapid clinical improvement of diabetes and hypertension [19]. In a case series of three patients with advanced ACC, the addition of metyrapone to EDP-M (etoposide, doxorubicin and cisplatin plus mitotane) regimen was effective and well tolerated [20]. This last study shows another possible indication for anti-cortisol treatment, in preparation for chemotherapy to reduce the risk of infections.

When mitotane reaches therapeutic plasma concentrations, the dose of other steroidogenesis inhibitors can be reduced, according to clinical response and cortisol levels. Later, glucocorticoid replacement therapy with cortisone acetate should be administered throughout treatment with mitotane.

Osilodrostat is the most recently introduced drug for the treatment of hypercortisolism, so the evidence in patients with ACC is limited. One study demonstrated the efficacy and safety of osilodrostat in patients with CS due to ACC [21]. The study showed that osilodrostat had quick action while sensitivity to the drug varied among the patients; for this reason, a wide range of doses was used and monitoring of serum and urinary cortisol 2–5 days after the beginning of treatment is recommended to adjust dosing. A block-and-replace therapy is suggested to enable the use of larger doses with the aim of ensuring more rapid efficacy while avoiding adrenal insufficiency [21].

The management of comorbidities of hypercortisolism is key for patients with ACC and CS. The monitoring of blood pressure and glycemic levels is necessary to optimize the treatment of hypertension and diabetes. After adrenalectomy or start of steroidogenesis inhibitors, serum cortisol should decrease, allowing the dose of antihypertensive and antidiabetic medications to be reduced.

As hypercortisolism induces immunosuppression, to prevent the occurrence of opportunistic infections, trimethoprim-sulfamethoxazole should be administered in patients with very high 24-h urinary free cortisol levels and/or additional risk factors (hyperglycemia, obesity) [22] since pneumonia due to *Pneumocystis jirovecii* in patients with moderate-severe hypercortisolism is associated with high mortality rate [23]. In the case of infections a prompt treatment is mandatory.

Thromboembolic events must be treated with anticoagulant therapy, and they can be prevented with pharmacologic (i.e., low-molecular-weight heparin) or non-pharmacologic (i.e., elastic compression stockings and intermittent pneumatic leg compression) prophylaxis, according to the individual risk of bleeding [22].

Acknowledgments This chapter is based upon work from COST Action CA20122 Harmonisation, supported by COST.

References

1. Puglisi S, Perotti P, Pia A, et al. Adrenocortical carcinoma with hypercortisolism. Endocrinol Metab Clin N Am. 2018;47(2):395–7.
2. Fassnacht M, Dekkers OM, Else T, et al. European Society of Endocrinology Clinical Practice Guidelines on the management of adrenocortical carcinoma in adults, in collaboration with the European Network for the Study of Adrenal Tumors. Eur J Endocrinol. 2018;179(4):G1–G46.
3. Puglisi S, Calabrese A, Ferraù F, et al. New findings on presentation and outcome of patients with adrenocortical cancer: results from a national cohort study. J Clin Endocrinol Metab. 2023;108(10):2517–25.
4. Nieman LK. Cushing's syndrome: update on signs, symptoms and biochemical screening. Eur J Endocrinol. 2015;173(4):M33–8.
5. Pivonello R, Isidori AM, De Martino MC, et al. Complications of Cushing's syndrome: state of the art. Lancet Diabetes Endocrinol. 2016;4(7):611–29.
6. Else T, Kim AC, Sabolch A, et al. Adrenocortical carcinoma. Endocr Rev. 2014;35(2):282–326.
7. Fassnacht M, Johanssen S, Quinkler M, et al. Limited prognostic value of the 2004 International Union Against Cancer staging classification for adrenocortical carcinoma: proposal for a Revised TNM Classification. Cancer. 2009;115(2):243–50.
8. Beuschlein F, Weigel J, Saeger W, et al. Major prognostic role of Ki67 in localized adrenocortical carcinoma after complete resection. J Clin Endocrinol Metab. 2015;100(3):841–9.
9. Calabrese A, Basile V, Puglisi S, et al. Adjuvant mitotane therapy is beneficial in non-metastatic adrenocortical carcinoma at high risk of recurrence. Eur J Endocrinol. 2019;180(6):387–96.
10. Elhassan YS, Altieri B, Berhane S, et al. S-GRAS score for prognostic classification of adrenocortical carcinoma: an international, multicenter ENSAT study. Eur J Endocrinol. 2021;186(1):25–36.
11. Berruti A, Terzolo M, Sperone P, et al. Etoposide, doxorubicin and cisplatin plus mitotane in the treatment of advanced adrenocortical carcinoma: a large prospective phase II trial. Endocr Relat Cancer. 2005;12(3):657–66.
12. Abiven G, Coste J, Groussin L, et al. Clinical and biological features in the prognosis of adrenocortical cancer: poor outcome of cortisol-secreting tumors in a series of 202 consecutive patients. J Clin Endocrinol Metab. 2006;91(7):2650–5.
13. Berruti A, Fassnacht M, Haak H, et al. Prognostic role of overt hypercortisolism in completely operated patients with adrenocortical cancer. Eur Urol. 2014;65(4):832–8.
14. Margonis GA, Kim Y, Tran TB, et al. Outcomes after resection of cortisol-secreting adrenocortical carcinoma. Am J Surg. 2016;211(6):1106–13.
15. Vanbrabant T, Fassnacht M, Assie G, Dekkers OM. Influence of hormonal functional status on survival in adrenocortical carcinoma: systematic review and meta-analysis. Eur J Endocrinol. 2018;179(6):429–36.
16. Zheng S, Cherniack AD, Dewal N, et al. Comprehensive pan-genomic characterization of adrenocortical carcinoma. Cancer Cancer Cell. 2016;29(5):723–36. s
17. Parikh PP, Rubio GA, Farra JC, Lew JI. Nationwide analysis of adrenocortical carcinoma reveals higher perioperative morbidity in functional tumors. Am J Surg. 2018;216(2):293–8.
18. Puglisi S, Calabrese A, Basile V, et al. New perspectives for mitotane treatment of adrenocortical carcinoma. Best Pract Res Clin Endocrinol Metab. 2020;34(3):101415.
19. Corcuff JB, Young J, Masquefa-Giraud P, et al. Rapid control of severe neoplastic hypercortisolism with metyrapone and ketoconazole. Eur J Endocrinol. 2015;172(4):473–81.
20. Claps M, Cerri S, Grisanti S, et al. Adding metyrapone to chemotherapy plus mitotane for Cushing's syndrome due to advanced adrenocortical carcinoma. Endocrine. 2018;61(1):169–72.

21. Tabarin A, Haissaguerre M, Lassole H, et al. Efficacy and tolerance of osilodrostat in patients with Cushing's syndrome due to adrenocortical carcinomas. Eur J Endocrinol. 2022;186(2):K1–4.
22. Varlamov EV, Langlois F, Vila G, Fleseriu M. Management of endocrine disease: Cardiovascular risk assessment, thromboembolism, and infection prevention in Cushing's syndrome: a practical approach. Eur J Endocrinol. 2021;184(5):R207–24.
23. van Halem K, Vrolijk L, Pereira AM, de Boer MGJ. Characteristics and mortality of Pneumocystis pneumonia in patients with Cushing's syndrome: a plea for timely initiation of chemoprophylaxis. Open Forum Infect Dis. 2017;4(1):ofx002.

Management of Hereditary Syndromes Associated with Pheochromocytoma/Paraganglioma

7

Mara Giacché

7.1 Introduction

Identification of an inherited pheochromocytoma/paraganglioma (PPGL) has important implications for the patient, as the germline mutation may have an impact on treatment in advanced disease, may influence surgical strategy (sparing surgery vs adrenalectomy) and drive postoperative care. Care of the patient with hereditary PPGL also includes taking care of the family members who test positive on genetic testing. Patients with PPGL syndrome and asymptomatic carriers of mutations in PPGL susceptibility genes need an extensive and lifelong surveillance program which considers their risk of relapse or development of new PPGL and the risk related to the involvement of other organs [1].

Surveillance programs should consider the biological characteristics of the specific gene mutation (risk of malignancy, penetrance), but also the possible radiological risk and psychological impact of such a long-term follow-up, especially in healthy mutation carriers.

Since these are rare neoplastic syndromes, some of which have only recently been identified, surveillance programs are not always evidence-based but are rather suggestions based on expert opinion and recommendations issued by consensus conferences.

M. Giacché (✉)
Internal Medicine, Department of Clinical and Experimental Sciences, University of Brescia at ASST Spedali Civili di Brescia, Brescia, Italy
e-mail: mara.giacche@asst-spedalicivili.it

© The Author(s) 2025
G. A. M. Tiberio (ed.), *Primary Adrenal Malignancies*, Updates in Surgery,
https://doi.org/10.1007/978-3-031-62301-1_7

7.2   Neurofibromatosis Type 1

Neurofibromatosis type 1 (NF1) is an autosomal dominant multisystem disease characterized by an extremely variable phenotype. The cutaneous phenotype presenting with multiple café-au-lait macules and cutaneous neurofibromas is most common and gives the disease its name. However, the clinical course of the disease is dependent on the involvement of other organs. The clinical diagnosis according to criteria developed by the United States National Institutes of Health (NIH) is based on the presence of at least two of the following [2]:

- six or more café-au-lait macules (>5 mm in diameter prepuberty, or > 15 mm postpuberty);
- at least 2 neurofibromas or 1 plexiform neurofibroma;
- axillary or inguinal freckling;
- optic glioma;
- two or more Lisch nodules;
- typical bony lesions (thickening of long bone cortex, pseudoarthrosis, sphenoid dysplasia);
- first degree relative with NF1.

The extent of skin manifestations has often a modest influence on the quality of life of patients, except in cases of severe esthetic impact; similarly, optic glioma and astrocytoma have generally a favorable outcome. The patients' quality of life is instead affected by the common behavioral disorders, specific learning disability and deep nodular neurofibromas or voluminous plexiform neurofibromas, which are responsible for neurologic deficits and severe neuropathic pain. Plexiform neurofibroma in 10% of cases may degenerate into malignant peripheral nerve sheath tumors: a rapid increase in volume, persistent pain, or exacerbation of existing pain with a change in consistency of the tumor are clues of malignant transformation. NF1 subjects have an increased risk of breast cancer and gastrointestinal stromal tumors (GIST). Surveillance in affected patients includes annual mammography from the age of 30 years and contrast-enhanced magnetic resonance imaging (MRI) in subjects with a family history of breast cancer. There are no clear indications for GIST surveillance: some authors propose abdominal ultrasound (US) every 24 months and annual complete blood count together with vitamin D and calcium metabolism assessment for the risk of osteoporosis. Currently, there is no consensus for PPGL screening in NF1 patients but, considering that affected subjects are often asymptomatic, biochemical assessment every 2 years starting at the age of 14 with 24-h urinary fractionated metanephrines could be a reasonable surveillance strategy and should also be considered prior to elective surgery and for women planning a pregnancy. Lifelong annual biochemical surveillance should be proposed for subjects with a prior diagnosis of PPGL [3].

Abdominal imaging (computed tomography [CT], MRI) and functional study with ($^{123/131}$I-MIBG scintigraphy, or ^{18}F-DOPA positron-emission tomography [PET]/CT) are indicated only if biochemistry is abnormal. NF1 mutated PPGLs belong to cluster 2 kinase signaling-related tumors (see Chap. 4); for these tumors ^{18}F-DOPA PET/CT is the preferred functional imaging modality due to the high uptake by the tumoral tissue compared to normal adrenal gland, allowing detection of multiple lesions within the adrenal parenchyma [4].

7.3 Multiple Endocrine Neoplasia 2

Multiple endocrine neoplasia 2 (MEN2) is a disease with autosomal dominant transmission, characterized by three clinical variants (MEN2a, MEN2b, and FMTC) as a result of the variable aggregation of medullary thyroid cancer (MTC), pheochromocytoma (PHEO), primary hyperparathyroidism (PHPT) and syndromic features (marfanoid habitus, mucosal neuromas, hindgut hypergangliosis, and thick corneal nerves) (Table 7.1).

The different clinical manifestations of MEN2 are often metachronous and not always expressed, MTC is highest penetrant and mostly affects morbidity and survival. An almost exclusive feature of *RET* gene mutations is the genotype-phenotype correlation, which allows prediction of the clinical expression for the different *RET* mutations. The American Thyroid Association (ATA) guidelines have classified mutations with the highest risk for MTC as HST (patients with *RET* codon M918T mutation, associated with MEN2B), mutations with a high risk as H (patients with C634F/G/R/S/W/Y and A883F mutations), and mutations with a moderate risk as MOD (Table 7.2).

Surveillance for *RET* mutation carriers is then strongly conditioned by the risk of medullary thyroid cancer, and includes indication for prophylactic thyroidectomy, which is timed according to the ATA risk classification [5].

Due to the high penetrance of aggressive MTC also at very young age, prophylactic thyroidectomy is recommended in the first year of life for MEN2B subjects (HST-risk mutation) and by at the age of 5 years or earlier, based on calcitonin level

Table 7.1 Clinical features of multiple endocrine neoplasia 2 (MEN2) syndromes

MEN2 variants	MTC	PHEO	PHPT	Non-endocrine features
MEN2A	95%	20–50%	20–30%	–
MEN2B	95%	50%	–	Marfanoid habitus, mucosal neuromas, hindgut hypergangliosis and thick corneal nerves, cutaneous lichen amyloidosis
FMTC	100%	–	–	–

MTC medullary thyroid cancer, *PHEO* pheochromocytoma, *PHPT* primary hyperparathyroidism

Table 7.2 Relationship between common *RET* mutations and the risk of aggressive MTC, PHEO and PHPT

RET mutation[a]	Risk of MTC[b]	Risk of PHEO	Risk of PHPT
G533C	MOD	10%	–
C609F/G/R/S/Y	MOD	10–20%	10%
C611F/G/S/Y/W	MOD	10–20%	10%
C618F/R/S	MOD	10–20%	10%
C620F/R/S	MOD	10–20%	10%
C630R/Y	MOD	10–20%	10%
D631Y	MOD	50%	–
C634F/G/R/S/W/Y	H	50%	20%
K666E	MOD	10%	–
E768D	MOD	–	–
L790F	MOD	10%	–
V804L	MOD	10%	10%
V804M	MOD	10%	10%
A883F	H	50%	–
S891A	MOD	10%	10%
R912P	MOD	–	–
M918T	HST	50%	–

MTC medullary thyroid cancer; *PHEO* pheochromocytoma; *PHPT* primary hyperparathyroidism
[a]*RET* codon mutations are indicated according to the one-letter code
[b]Risk of MTC is indicated according to the American Thyroid Association guidelines (*HST* highest, *H* high, *MOD* moderate). Modified from [5]

and neck ultrasound for MEN2A subjects (H-risk mutations). For subjects with MOD-risk mutations the timing of prophylactic thyroidectomy should be optimized according to the family history and patient's desires: follow-up involves annual neck US and serum calcitonin assay, with thyroidectomy being performed when the calcitonin level becomes elevated.

PHPT is usually mild and asymptomatic and occurs in 20–30% of MEN2A patients, mostly with mutation in codon 634.

Surveillance for PPGL with annual 24-h urinary fractionated metanephrines should start at the age of 11 years for HST-risk and H-risk mutations and at the age of 16 years for MOD-risk mutations; abdominal imaging (CT, MRI) is indicated only in the event of abnormal biochemistry. For a functional study, considering that *RET*-mutated PHEOs belong to cluster 2 kinase signaling-related tumors, ^{18}F-DOPA PET/CT is considered the first choice in preference to the traditionally used [123/131]I-MIBG scintigraphy. The second suggested functional imaging study is probably ^{68}Ga-DOTA PET/CT.

7.4 Von Hippel Lindau Syndrome

Von Hippel Lindau (VHL) syndrome is a hereditary neoplastic disease with autosomal dominant transmission. It is characterized by multiorgan involvement with predisposition to benign and malignant neoplasia; the tumors which mostly impact the clinical course of disease are central nervous system and retinal hemangioblastomas, clear cell carcinoma, neuroendocrine tumors and PPGL.

VHL syndrome was in the past classified into type 1 and type 2, according to the risk of developing PHEO: lower for type 1 (<10%) and higher (40–60%) for VHL type 2. VHL type 2 was further classified into subtype 2A (lower risk of renal cancer), subtype 2B (high risk of renal cancer), and subtype 2C with only risk for PHEO without other manifestations of VHL syndrome. Application of genetic screening in clinical practice has excluded the presence of a genotype/phenotype association, so the patient's possible attribution to a clinical phenotype should never influence the follow-up. Surveillance programs should consider all possible organ involvement, with the aim of achieving a preclinical diagnosis and early treatment of lesions. The surveillance protocol for *VHL*-mutated subjects is summarized in Table 7.3.

Surveillance for PPGL should start at the age of 4 years considering that *VHL* mutations are often involved in pediatric PHEOs (about 40%) and consists of annual blood pressure monitoring and plasma-free metanephrine, normetanephrine, 3-methoxytyramine and/or measurement of 24-h urinary metabolite excretion. Annual abdominal US and MRI every 2 years are suggested also for early detection of kidney cancer and neuroendocrine tumors. Functional studies are indicated only in the case of abnormal biochemistry and/or suspected abnormality at imaging: *VHL*-related PPGL have lower expression of somatostatin receptor 2 (SSTR2), so the most sensitive functional imaging study is probably ^{18}F-DOPA PET/CT where

Table 7.3 Surveillance protocol for von Hippel-Lindau disease

Central nervous system hemangiomas	Retinal hemangiomas	Pheochromocytoma	Renal cell cancer and neuroendocrine tumors	Endolymphatic sac tumors
MRI brain and full spine	Ophthalmic examinations	Blood pressure monitoring and biochemistry[a]	Abdominal ultrasound	Audiogram
Every 1–2 years	Every 1 year	Every 1 year	Every 1 year	Every 2–3 years
			Abdominal MRI Every 2 years	MRI of the internal auditory canal in subjects with repeat ear infections
From the age of 16	From the age of 1	From the age of 4	From the age of 12	From the age of 16

[a]Biochemistry: plasma free normetanephrine, metanephrine, 3-methoxytyramine, and/or measurement of 24-h urinary metabolite excretion

available; otherwise, [123/131]I-MIBG scintigraphy and, as a second choice, ^{68}Ga-DOTA PET/CT can be used.

7.5   *SDHx*-associated Hereditary PPGL

PPGL is the most common and characteristic expression of *SDHx* mutations, but mutations in these genes also predispose to renal cancer and to wild GIST (GIST without mutation in the *KIT* and *PDGFRA* genes). The surveillance protocol for *SDHx* mutations is not easy to devise: in fact, if on the one hand the risk of malignancy associated with *SDHB*, *SDHD* and probably *SDHA* mutations would warrant an intensive surveillance protocol, on the other hand the low penetrance of *SDHB* and *SDHA* mutations could suggest lengthening the intervals between follow-up assessments. Follow-up for gene mutation carriers comprises annual blood pressure measurement and physical examination, plasma free metanephrine, normetanephrine, 3-methoxytyramine and/or measurement of the 24-h urinary metabolite excretion, and MRI from the base of the skull to the pelvis every 2 or 3 years. Intensity of screening should be stronger for patients with a history of PPGL: MRI is recommended every 2 years. Whole-body MRI should be replaced with MRI of the skull base, neck, abdomen and pelvis, and low-dose chest CT.

SDHx-related PPGLs intensely express the SSTR2, so ^{68}Ga-DOTATATE PET/CT is considered the most sensitive functional imaging test. This could be performed at initial screening and subsequently only if abnormal findings are detected at biochemistry or MRI; there is no consensus on the hypothesis of alternating the MRI study with ^{68}Ga-DOTATATE PET/CT every 2–3 years. The surveillance protocol for VHL-mutated subjects is summarized in Table 7.4.

Table 7.4 Surveillance protocol for *SDHx* mutation carriers

Age to begin surveillance	*SDHB*: From the age of 6–10 Other *SDHx*: From the age of 10–15
Physical examination and blood pressure check	Every 1 year
Biochemistry[a]	Every 1 year
MRI (base of the skull to pelvis)	Every 2–3 years
^{68}Ga-SSA PET	Initial screening in adultThen only if abnormal biochemistry/imaging

[a]Biochemistry: plasma free normetanephrine, metanephrine, 3-methoxytyramine, and/or measurement of 24-h urinary metabolite excretion

7.6 *TMEM127*- and *MAX*-associated PPGL

Follow-up surveillance for *TMEM127* and *MAX* gene mutations involves annual physical examination, blood pressure measurement, annual plasma free metanephrine, normetanephrine, 3-methoxytyramine and/or measurement of the 24-h urinary metabolite excretion. Abdominal MRI could be performed every 2–3 years.

7.7 *FH*-associated PPGL

FH gene mutations are responsible for hereditary leiomyomatosis and papillary renal cell carcinoma (HLRCC), a complex neoplastic syndrome with autosomal dominant transmission, characterized by predisposition to cutaneous and uterine leiomyomata and papillary kidney cancer. PPGL have been described in only a small number of families. Often massive uterine leiomyomatosis causes severe bleeding and hysterectomy is necessary. Penetrance for renal cancer is fairly low (20%), and usually the cancer is a solitary but rapidly aggressive lesion. Also *FH*-PPGL may have an aggressive behavior [6]. Due to the small number of families studied until now, penetrance for PPGL is unknown, but is probably low.

Mutation carriers need annual gynecologic evaluation from the age of 20 years and abdominal annual MRI from the age of 8 years [7]. Biochemical screening for PPGL could probably start in young adulthood (after 18 years).

References

1. Plouin PF, Amar L, Dekkers OM, et al. European Society of Endocrinology Clinical Practice Guideline for long-term follow-up of patients operated on for a phaeochromocytoma or a paraganglioma. Eur J Endocrinol. 2016;174(5):G1–G10.
2. Legius E, Messiaen L, Wolkenstein P, et al. Revised diagnostic criteria for neurofibromatosis type 1 and Legius syndrome: an international consensus recommendation. Genet Med. 2021;23(8):1506–13.
3. Gruber LM, Erickson D, Babovic-Vuksanovic D, et al. Pheochromocytoma and paraganglioma in patients with neurofibromatosis type 1. Clin Endocrinol. 2017;86(1):141–9.
4. Nölting S, Bechmann N, Taieb D, et al. Personalized management of pheochromocytoma and paraganglioma. Endocr Rev. 2022;43(2):199–239. Erratum in: Endocr Rev. 2022;43(2):440;437–9
5. Wells SA Jr, Asa SL, Dralle H, et al. Revised American Thyroid Association guidelines for the management of medullary thyroid carcinoma. Thyroid. 2015;25(6):567–610.
6. Castro-Vega LJ, Buffet A, De Cubas AA, et al. Germline mutations in FH confer predisposition to malignant pheochromocytomas and paragangliomas. Hum Mol Genet. 2014;23(9):2440–6.
7. Schultz KAP, Rednam SP, Kamihara J, et al. PTEN, DICER1, FH, and their associated tumor susceptibility syndromes: clinical features, genetics, and surveillance recommendations in childhood. Clin Cancer Res. 2017;23(12):e76–82.

Adrenal Incidentaloma

Anna Maria Elena Perini, Antonio Gigante, Soraya Puglisi,
Laura Saba, and Massimo Terzolo

8.1 What Does "Adrenal Incidentaloma" Mean?

Adrenal incidentalomas are adrenal masses discovered by chance while performing imaging studies for reasons other than the suspicion or follow-up of adrenal diseases [1]. *Per se*, the term adrenal incidentaloma does not identify a precise diagnosis; rather, it is an umbrella definition encompassing many different tumor types.

8.2 How Often Will I Find an Adrenal Incidentaloma in My Practice?

According to the most recent radiological studies [2–4], the frequency of adrenal masses ranges from 1.4% to 7.3% in the general population. Similarly, autopsy studies reported a prevalence of adrenal masses of about 2%, ranging from 1% to 8.7% [1]. This variability is partially due to the differences in the age of the selected populations, since the prevalence of adrenal incidentalomas increases with age, approximating 10% in patients older than 80 years. However, it should be kept in mind that, although adrenal tumors are uncommon in young people and very rare in children and adolescents, in these patients the masses are more frequently hormone-secreting and malignant [1]. Therefore, the first lesson to draw from these epidemiological data is that in most cases we will manage elderly patients, often affected by several comorbid conditions, while in few cases we will deal with patients <40 years of age, but this context should ring an alarm bell in our mind.

A. M. E. Perini · A. Gigante · S. Puglisi (✉) · L. Saba · M. Terzolo
Internal Medicine, Department of Clinical and Biological Sciences, University of Turin, San Luigi Gonzaga University Hospital, Orbassano (Turin), Italy
e-mail: annamariaelena.perini@unito.it; antonio.gigante@unito.it; soraya.puglisi@unito.it; laura.saba@unito.it; massimo.terzolo@unito.it

© The Author(s) 2025
G. A. M. Tiberio (ed.), *Primary Adrenal Malignancies*, Updates in Surgery,
https://doi.org/10.1007/978-3-031-62301-1_8

8.3   What Type of Adrenal Tumor Can Be Found Incidentally?

Adrenal masses can be classified as follows [5]:

1. Adrenal adenomas and macronodular bilateral adrenal hyperplasia;
2. Other benign lesions (myelolipomas, cysts, hematomas, other);
3. Adrenocortical carcinomas (ACC);
4. Other malignant tumors (metastases, sarcomas, lymphoma);
5. Pheochromocytomas.

The majority of these lesions are benign (Table 8.1) [1, 6]; however, malignant or hormonally active adrenal lesions are associated with poor prognosis if not promptly identified and correctly treated. For this reason, determining whether the discovered mass has the potential to be malignant is of paramount importance [1].

Table 8.1 Comparison between frequencies of adrenal incidentaloma types based on clinical studies (hormonal and radiological work-up) and surgical studies (patients who underwent adrenalectomy)

Etiology	Frequency	
	Clinical studies[a]	Surgical studies[b]
Adenoma	80–85%	49–69%
Non-functioning	40–70%	52–75%
MACS	20–50%	1–15%
Aldosterone-secreting	2–5%	2–7%
Overt Cushing's syndrome	1–4%	n.a.
Pheochromocytoma	1–5%	11–23%
Carcinoma	0.4–4%	1.2–12%
Metastasis	3–7%	0–21%
Myelolipoma	3–7%	7–15%
Cyst and pseudocyst	1%	4–22%
Ganglioneuroma	1%	0–8%

MACS mild autonomous cortisol secretion; *n.a.* not available
[a]Data from [1]
[b]Data from [6]

8.4 What Should I Do Next After Discovering an Adrenal Incidentaloma?

According to the guidelines of the European Society of Endocrinology (ESE) and European Network for the Study of Adrenal Tumors (ENSAT) [1], when an adrenal mass is discovered, clinicians should evaluate and define:

1. Risk of malignancy;
2. Hormonal activity.

The risk of malignancy and the hormonal excess should be evaluated simultaneously.

8.4.1 Risk of Malignancy

Malignancy is reported in 5–8% of patients with adrenal incidentalomas [5]. In a population-based study, the most common etiology in the category of malignant adrenal tumors was adrenal metastases (86%), while only 3.6% were ACC [3]. Nonetheless, ACC is the most frequent malignant adrenal tumor reported in endocrine case series and recent studies demonstrated that 38–44% of ACC present as incidentalomas [7, 8].

The malignant potential is influenced by both (a) patient and (b) tumor characteristics. Since no parameter can confirm or exclude by itself the malignant nature of the lesion, the evaluation should consider globally the following features.

(a) *Features influencing the "a priori" risk of malignancy* [5]
   - Young age: adrenal incidentalomas are uncommon in children and adolescents and usually show an increased malignant potential in this population.
   - Constitutional symptoms: low-grade fever, fatigue, weight loss.
   - History of extra-adrenal malignancies or genetic syndromes associated with increased cancer risk (ACC or pheochromocytoma).
(b) *Imaging characteristics of the lesion* [1]
   Non-contrast computed tomography (CT) is recommended as the first imaging modality, if not yet performed. This examination provides data about size, lipid content and homogeneity of the lesion. A key element is the evaluation of the mass density expressed in Hounsfield units (HU), because low HU values reflect high lipid content, and benign lesions are usually homogeneous and lipid-rich. Combining the characteristics of size, density and homogeneity, the ESE/ENSAT guidelines summarize the recommendations for different scenarios:
   - The adrenal mass is homogenous and ≤ 10 HU: benign lesion, no additional imaging is required after exclusion of hormonal excess.
   - Tumor size ≥4 cm and HU >20, or mass is heterogeneous: a malignant lesion should be suspected. In these cases, surgery is usually recommended after completing staging procedures (i.e., chest CT or ^{18}F-FDG positron emission tomography [PET]/CT).

- Cases that do not fit any of the previous categories need a multidisciplinary approach with an expert team. The team should consist of staff qualified for the management of adrenal tumors, and should comprise a radiologist, an endocrinologist and a surgeon. The options include proceeding immediately to another imaging test (i.e., FDG PET/CT or MRI) or follow-up imaging after 3–6 months, or proceeding swiftly to surgery, and the choice depends on patient's age, history, clinical presentation, and imaging characteristics.

8.4.2 Hormonal Activity

Based on expert consensus, the evaluation of hormonal activity is indicated when adrenal tumor size is at least 1 cm or in the presence of clinical signs and symptoms suggestive for hormonal excess [1].

Studies show that up to 30–50% of adrenal lesions are responsible for hormone excess, often associated with increased cardiometabolic morbidity and mortality [2, 9–13].

A careful history should be collected and a physical examination focused on potential signs of overt hormone excess should be performed in all patients.

The evaluation of hormonal activity includes: (a) cortisol; (b) free plasma or urinary fractionated metanephrines; (c) aldosterone/renin ratio; (d) sex steroids and precursors of steroidogenesis.

(a) *Cortisol*

Adrenal cortisol secretion is the most frequent finding; therefore, it is mandatory to exclude autonomous cortisol secretion using the 1-mg overnight dexamethasone suppression test (DEX) [1, 6, 14]. This test assesses the normal function of the hypothalamus-pituitary-adrenal (HPA) axis feedback. Serum cortisol is evaluated in the morning at 8.00 am, after the patient has taken 1 mg of dexamethasone between 11.00 p.m. and midnight the previous night. A value of ≤50 nmol/L (≤ 1.8 μg/dL) may be regarded as physiologic, excluding cortisol excess and reflecting normal HPA axis suppression. Recently, the prevalent opinion is that DEX results should be considered as a continuous rather than a categorical variable; however, defining a cut-off is useful for distinguishing between functioning and non-functioning adenomas. Values above 50 nmol/L should be considered indicative of mild autonomous cortisol secretion (MACS). In these cases, the diagnosis should be confirmed with a second DEX test, while additional hormonal tests (late night salivary cortisol, 24-h urinary free cortisol) may be required depending on clinical circumstances, and pituitary disease should be excluded with ACTH level measurement. MACS is not associated with an increased risk of developing an overt Cushing's syndrome (<1%) [15, 16], but the data have shown that it is associated with increased metabolic and cardiovascular risks [9–12]. In particular, MACS is associated with increased all-cause mortality, especially in women younger than 65 years [13]. Moreover, some but not all studies found a higher prevalence of osteoporosis and asymptomatic vertebral fractures in patients with MACS [17–19], particularly in postmenopausal women [20]. Screening for hypertension, type 2 diabetes mellitus and vertebral fractures are indicated in patients with MACS [1].

(b) *Free plasma or urinary fractionated metanephrines*

When discovering an adrenal lesion with indeterminate imaging characteristics or in the case of genetic syndromes harboring an increased risk of pheochromocytoma, clinicians should exclude this diagnosis through the measurement of free plasma or urinary fractionated metanephrines. This evaluation is not required when the mass has <10 HU.

Before any surgery or biopsy, pheochromocytoma should be excluded to prevent a catecholamine crisis and define the best preparation and intraoperative management for the patient.

(c) *Aldosterone/renin ratio*

This test is indicated for patients with hypertension or unexplained hypokalemia to exclude primary aldosteronism.

(d) *Sex steroids and precursors of steroidogenesis*

These measurements are indicated in patients whose clinical and imaging characteristics are suggestive for ACC.

Adrenal biopsy should be considered to exclude an adrenal metastasis when this influences patient management. Moreover, it may be useful when rare tumors like adrenal lymphoma or sarcoma are suspected. Otherwise, adrenal biopsy is not part of the standard diagnostic workup of adrenal incidentalomas, because it has low diagnostic accuracy, especially for ACC, and is burdened by possible complications [1, 21]. Before performing an adrenal biopsy, catecholamine excess should always be excluded to avoid cardiovascular crises during the procedure [1].

8.5 Which Treatment?

The decision whether to perform an adrenalectomy or to simply follow up the adrenal incidentaloma over time with clinical, hormonal and imaging assessments should be guided by the patient's characteristics (i.e.: performance status, age, patient's preference), the malignant potential and the hormonal activity of the adrenal lesion.

It is of crucial importance that a multidisciplinary expert team evaluates whether or not there is an indication to perform the adrenalectomy [1].

After surgery, patients with DEX test results ≥50 nmol/L are at risk of developing adrenal insufficiency. For this reason, these patients should undergo perioperative glucocorticoid treatment at surgical stress doses and, after surgery, they should be followed by an endocrinologist until recovery of the HPA axis has been documented [1].

8.6   Which Patients Deserve Particular Consideration?

8.6.1   Mild Autonomous Cortisol Secretion

In patients with MACS the indication for surgery should be individualized. Many factors should be taken into account, but age and comorbidities are the main ones. Older people show greater cortisol levels after DEX regardless of comorbidities, and there is evidence that the clinical significance of MACS decreases in patients older than 65 years [13]. For these reasons conservative management is the most frequent choice in older people. The presence of uncontrolled hypertension, diabetes or fragility fractures and evidence of progressive disease, associated with inappropriate end-organ damage, are the features clinicians should consider when evaluating the indication for surgery.

8.6.2   Bilateral Adrenal Incidentalomas

The initial evaluation of bilateral adrenal lesions is the same as used for unilateral one. Bilateral disease can be attributed to:

- Bilateral macronodular hyperplasia (congenital adrenal hyperplasia should be excluded by measuring 17-hydroxyprogesterone);
- Bilateral adrenal adenomas;
- Morphologically similar adrenal masses (non-adenoma);
- Morphologically different adrenal masses.

Bilateral incidentalomas (especially bilateral macronodular hyperplasia and bilateral adrenal adenomas) are more frequently associated with MACS and both adrenal glands can contribute to cortisol excess. In these patients, surgical management should be individualized and bilateral adrenalectomy should be reserved for those with Cushing's syndrome due to the high morbidity burden of this procedure.

In patients with large and bilateral metastases replacing the adrenal gland tissue, there is an increased risk of adrenal insufficiency, to be excluded with morning serum cortisol measurement.

8.6.3   Younger People (< 40 Years)

The approach to this population should be more aggressive due to increased risk of malignancy. For this reason, indeterminate adrenal masses should undergo surgical treatment [1].

Acknowledgments This chapter is based upon work from COST Action CA20122 Harmonisation, supported by COST.

References

1. Fassnacht M, Tsagarakis S, Terzolo M, et al. European Society of Endocrinology clinical practice guidelines on the management of adrenal incidentalomas, in collaboration with the European Network for the Study of Adrenal Tumors. Eur J Endocrinol. 2023;189(1):G1–G42.
2. Reimondo G, Castellano E, Grosso M, et al. Adrenal incidentalomas are tied to increased risk of diabetes: findings from a prospective study. J Clin Endocrinol Metab. 2020;105(4):dgz284.
3. Ebbehoj A, Li D, Kaur RJ, et al. Epidemiology of adrenal tumours in Olmsted County, Minnesota, USA: a population-based cohort study. Lancet Diabetes Endocrinol. 2020;8(11):894–902.
4. Jing Y, Hu J, Luo R, et al. Prevalence and characteristics of adrenal tumors in an unselected screening population: a cross-sectional study. Ann Intern Med. 2022;175(10):1383–91.
5. Bancos I, Prete A. Approach to the patient with adrenal incidentaloma. J Clin Endocrinol Metab. 2021;106(11):3331–53.
6. Terzolo M, Stigliano A, Chiodini I, et al. AME position statement on adrenal incidentaloma. Eur J Endocrinol. 2011;164(6):851–70.
7. Bancos I, Taylor AE, Chortis V, et al. Urine steroid metabolomics for the differential diagnosis of adrenal incidentalomas in the EURINE-ACT study: a prospective test validation study. Lancet Diabetes Endocrinol. 2020;8(9):773–81.
8. Puglisi S, Calabrese A, Ferraù F, et al. New findings on presentation and outcome of patients with adrenocortical cancer: results from a national cohort study. J Clin Endocrinol Metab. 2023;108(10):2517–25.
9. Di Dalmazi G, Vicennati V, Garelli S, et al. Cardiovascular events and mortality in patients with adrenal incidentalomas that are either non-secreting or associated with intermediate phenotype or subclinical Cushing's syndrome: a 15–year retrospective study. Lancet Diabetes Endocrinol. 2014;2(5):396–405.
10. Debono M, Bradburn M, Bull M, et al. Cortisol as a marker for increased mortality in patients with incidental adrenocortical adenomas. J Clin Endocrinol Metab. 2014;99(12):4462–70.
11. Patrova J, Kjellman M, Wahrenberg H, Falhammar H. Increased mortality in patients with adrenal incidentalomas and autonomous cortisol secretion: a 13-year retrospective study from one center. Endocrine. 2017;58(2):267–75.
12. Kjellbom A, Lindgren O, Puvaneswaralingam S, et al. Association between mortality and levels of autonomous cortisol secretion by adrenal incidentalomas: a cohort study. Ann Intern Med. 2021;174(8):1041–9.
13. Deutschbein T, Reimondo G, Di Dalmazi G, et al. Age-dependent and sex-dependent disparity in mortality in patients with adrenal incidentalomas and autonomous cortisol secretion: an international, retrospective, cohort study. Lancet Diabetes Endocrinol. 2022;10(7):499–508.
14. Reimondo G, Puglisi S, Pia A, Terzolo M. Autonomous hypercortisolism: definition and clinical implications. Minerva Endocrinol. 2019;44(1):33–42.
15. Elhassan YS, Alahdab F, Prete A, et al. Natural history of adrenal incidentalomas with and without mild autonomous cortisol excess: a systematic review and meta-analysis. Ann Intern Med. 2019;171(2):107–16.
16. Reimondo G, Muller A, Ingargiola E, et al. Is follow-up of adrenal incidentalomas always mandatory? Endocrinol Metab (Seoul). 2020;35(1):26–35.
17. Chiodini I, Morelli V, Masserini B, et al. Bone mineral density, prevalence of vertebral fractures, and bone quality in patients with adrenal incidentalomas with and without subclinical hypercortisolism: an Italian multicenter study. J Clin Endocrinol Metab. 2009;94(9):3207–14.
18. Morelli V, Eller-Vainicher C, Salcuni AS, et al. Risk of new vertebral fractures in patients with adrenal incidentaloma with and without subclinical hypercortisolism: a multicenter longitudinal study. J Bone Miner Res. 2011;26(8):1816–21.
19. Favero V, Eller-Vainicher C, Morelli V, et al. Increased risk of vertebral fractures in patients with mild autonomous cortisol secretion. J Clin Endocrinol Metab. 2024;109(2):e623–32.

20. Zavatta G, Vicennati V, Altieri P, et al. Mild autonomous cortisol secretion in adrenal incidentalomas and risk of fragility fractures: a large cross-sectional study. Eur J Endocrinol. 2023;188(4):343–52.
21. Bancos I, Tamhane S, Shah M, et al. Diagnosis of endocrine disease: The diagnostic performance of adrenal biopsy: a systematic review and meta-analysis. Eur J Endocrinol. 2016;175(2):R65–80.

Surgery for Adrenocortical Carcinoma

9

Guido A. M. Tiberio, Silvia Ministrini, Giovanni Casole, Giacomo Gaverini, and Stefano M. Giulini

9.1   Introduction

A correct adrenalectomy represents the most important prognostic determinant of the clinical course of a patient with adrenocortical carcinoma (ACC) and it should be considered a prerequisite for cure. The definition of curative adrenalectomy is intriguing, as it has not yet been fully delineated. Multiple elements contribute to its achievement; among them, the integrity of the tumor during surgical manipulation, the extent of periadrenal soft-tissue clearance, the role of lymphadenectomy and the surgeon's expertise. The quality indicator for this surgery is represented by the rate of local/peritoneal recurrence, which is impressive also for early-stage tumors. Attention to every single detail leads to optimization of the surgical treatment of the disease. In this chapter we will describe how a correct adrenalectomy for cancer should be performed, and discuss the surgical strategy for recurrent local or peritoneal ACC.

9.2   Upfront Adrenalectomy: The Guidelines

Among the different guidelines issued in recent years [1–4], those published in 2018 by the European Society of Endocrinology in collaboration with the European Network for the Study of Adrenal Tumors [1] stand out for their completeness and reasoned development.

G. A. M. Tiberio (✉) · S. Ministrini · G. Casole · G. Gaverini · S. M. Giulini
General Surgery, Department of Clinical and Experimental Sciences, University of Brescia at ASST Spedali Civili di Brescia, Brescia, Italy
e-mail: guido.tiberio@unibs.it; ministrini.silvia@me.com; giovannicasole@gmail.com; giacomogave@alice.it; stefano.giulini@unibs.it

© The Author(s) 2025
G. A. M. Tiberio (ed.), *Primary Adrenal Malignancies*, Updates in Surgery,
https://doi.org/10.1007/978-3-031-62301-1_9

Concerning the surgical management of localized ACC, they state [1]:

"R.3.1. We recommend that adrenal surgery for suspected/confirmed ACC should be performed only by surgeons experienced in adrenal and oncological surgery";

"R.3.2. We recommend complete en bloc resection of all adrenal tumors suspected to be ACC including the peritumoral/periadrenal retroperitoneal fat. We recommend against enucleation and partial adrenal resection for suspected ACC. If adjacent organs are suspected to be invaded, we recommend en bloc resection. However, we suggest against the routine resection of the ipsilateral kidney in absence of direct renal invasion";

"R.3.4. We suggest that routine locoregional lymphadenectomy should be performed with adrenalectomy for highly suspected or proven ACC. It should include (as a minimum) the periadrenal and renal hilum nodes. All suspicious or enlarged lymph nodes identified on preoperative imaging or intraoperatively should be removed";

"R.3.7. We recommend perioperative hydrocortisone replacement in all patients with hypercortisolism that undergo surgery for ACC".

9.2.1   The Literature

Because of the rarity of ACC no prospective randomized studies exist and the literature offers only retrospective, often multicentric studies.

9.2.1.1   R.3.1

Adrenal surgery has low mortality and morbidity rates. A review of 9820 procedures by the French Association of Endocrine Surgeons reported a 30- and 90-day mortality rate of 0.8% and 1.5%, respectively. Mortality was unevenly distributed according to the hospital case-load: 1% vs. 0.4% (30-day) and 1.8% vs. 0.9% (90-day) in low- and high-volume institutions, respectively. At multivariate analysis the hospital case-load was independently associated with operative mortality (OR: 1.8, $p < 0.010$), along with other risk factors such as age, comorbidity, malignancy, open surgery and reintervention, and this was highly appreciable in high-risk patients ($p = 0.003$); a case-load of 32 patients/year was indicated as the best cut-off to recognize high-volume hospitals [5]. Similarly, considering a range of short-term indicators, a volume-outcome effect was recognized in England [6]; the benefits of centralization became appreciable above 10 procedures per year, with greater advantages for patients at the threshold of 20 adrenalectomies/surgeon/year and 30 per institution. In the Netherlands, the centralization of adrenal surgery led to an impressive improvement of survival: from 78% to 93% and from 42% to 63% after 1 and 5 years, respectively, for patients receiving potentially curative adrenalectomies (stage I–III) [7]. Panelists of the ESE-ENSAT guidelines [1] suggest a minimum of >20 adrenalectomies/year for the surgical treatment of primary malignancies; these should be performed only in referral institutions with dedicated adrenal tumor boards, considering the complexity of this surgery. These requirements also apply to those cases in which the clinical diagnosis of malignancy is not certain but deemed possible. Dedicated mentorship should be pursued by those interested in adrenal surgery.

9.2.1.2   R.3.2 and R.3.4

Curative surgery is a prerequisite for cure, and for this reason the surgeon should guarantee the best procedure, considering all the different factors that measure the quality of adrenal surgery.

It has long been appreciated that tumor effraction and positive margins negatively affect oncologic outcomes, favoring local/peritoneal recurrence, an expression of tumor seeding and tumor persistence. Tumor integrity and negative resection margins concur in the definition of curative adrenalectomy. Case series of patients operated on during the first decade of this century reported 5-year recurrence-free survival rates around 30% and 15%, and 5-year overall survival rates of 60–95% vs. 15–65% in the case of negative or positive margins, respectively [8–10]. Removal of the adrenal gland en bloc with the renal capsule and retroperitoneal connective tissue reduces the risk of tumor rupture and increases the rate of margin-negative resections. Margin status has such a relevant impact on survival that the panelists of the ESE-ENSAT guidelines recommend: "R.3.6. If the first surgery was suboptimal and macroscopically incomplete (R2 resection), we suggest to discuss repeat surgery in a multidisciplinary expert team". Surgery should be indicated if the residual tumor is detectable at cross-sectional imaging, and a curative R0 surgery is deemed possible.

Lymph node involvement affects local recurrence and survival. In recent series nodal metastases were reported in about one-third of cases subjected to lymphadenectomy [9, 11–13]. Despite this, lymphadenectomy is performed in 20–30% of cases and almost exclusively in academic referral centers. The reason for this is the lack of a clear, widely adopted and shared anatomically based surgical strategy dictating the dissection rules to which surgeons should adhere. In fact, a thorough evaluation of different series shows that the extent of lymphadenectomy is widely inhomogeneous, ranging from systematic removal of locoregional nodes to nodal sampling. Experts agree on a minimum number of 4–5 nodes to define a real lymphadenectomy [11, 12]. There is also consensus on considering the right paracaval and left para-aortic nodes, together with those at the ipsilateral renal hilum as first-level lymphatic stations. A clear description of the attitude toward lymphadenectomy in ACC was provided by the German ACC study Group: 47/283 stage I–III patients (16.6%) received lymphadenectomy, which was more likely to be performed in the case of larger tumors, stage III tumors and multivisceral resections. Median follow-up was 59 months for the nodal dissection group and 39 months for the non-dissection group; at these time points recurrence and disease-specific death rates were similar (68.1 vs. 60.6 and 29.8 vs. 30.5, respectively) between the groups, but time to recurrence was longer after lymphadenectomy (20.1 vs.12.8 months); the presence of lymphatic metastases had a negative prognostic role as disease-free and disease-specific survival were both shorter (12.5 vs. 31.3 months, $p = 0.058$, and 86.4 vs. 135 months, $p = 0.002$, respectively). Interestingly, considering the node-positive patients, after adjustment for age, tumor stage, multivisceral resection, adjuvant treatment and nodal status on preoperative imaging, lymphadenectomy resulted in a significant reduction of the risk of recurrence (HR 0.65; 95% CI 0.43–0.98, $p = 0.42$) and of disease-specific death (HR 0.54; 95% CI 0.29–0.99, $p = 0.049$), and this was particularly evident for stage III ACC. Gerry et al. confirmed the German Group's oncologic outcomes and detailed how lymphadenectomy has a limited impact on the postoperative course [14]. A third study based on 386 patients from the National Cancer Database [15] reached substantially similar results but also showed that the strongest prognostic factor for poor survival was the number of positive nodes, which showed a progressive effect: the hazard ratio was 2.3 (CI 1.5–3.6) for 1 node, 3.0 (CI 1.1–8.0) for 3 nodes, and 4 (CI 2.5–6.2) for ≥4 nodes. A different analysis of the same database [13] added some more information: lymph nodes were retrieved more frequently in open than in minimally invasive surgery, and the percentage of nodal metastases was proportional to the number of nodes examined. On these bases it is now clear that lymphadenectomy should complete any adrenalectomy for ACC. Interestingly, however, the panelists of the ESE-ENSAT guidelines [1] do not support surgical radicalization if a margin-negative adrenalectomy has been performed, as in the case of misdiagnosis at preoperative work-up; they consider the harm/benefit balance to be uncertain and prefer to support the early start of adjuvant therapy.

It is important to bear in mind that nephrectomy to facilitate the lymphadenectomy or to reach a "safer and easier margin" is abusive [16]. Nephrectomy is only admitted in the case of direct infiltration of the renal parenchyma, and, if this is minimal, nephron-sparing strategies should be considered. In fact, a preserved renal function may have an important role during the subsequent course of the disease, facilitating systemic treatments. There is no demonstration of a positive impact from the resection of neighboring organs (spleen, pancreas, colon, stomach, liver) if these are not infiltrated; indeed, these procedures increase the risk of complications and mortality.

9.2.1.3   R.3.7

Intra- and postoperative glucocorticoid replacement must be provided to all patients with adrenal autonomous cortisol secretion at the doses suggested for major stress by the guidelines [17]. The surgeon must consider that glucocorticoid deficiency also affects the coagulation cascade and may cause delayed postoperative hemorrhages. Since a critical phase for such deficiency is typically observed at the switch from parenteral to oral administration, it is safe to overlap the two routes for some days.

9.2.2   Reasoning to Establish a Sound Surgical Strategy

As for other organs, a curative resection of ACC should remove the entire area of embryonal development, deploying all efforts to maintain intact its envelope, and sectioning all vascular, lymphatic and nervous structures at their origin/confluence. This is the rationale at the basis of a curative D2 gastrectomy, of a total mesorectal excision for rectal cancer, of a complete mesocolic excision for colonic cancer or radical antegrade modular pancreatosplenectomy for pancreatic cancer. However, a major difference exists: the absence of a complete mesothelial envelope to define the surgical target. The complexity of the adrenal vascular and lymphatic system reflects the gland's peculiar embryogenesis, both ectodermal (medulla) and mesodermal (cortical). Accordingly, embryonal development follows two different axes: horizontal, from the celiac ganglia, and ascendant, from the renal hilum. Since the two embryological components share the vascular and lymphatic systems, the extension of the area to be dissected is relatively large. The lymphatic system is mainly posterior, toward the celiac trunk and renal hilum and finds its medial limit along the right or left side of the aorta with prolongation up to the renocaval confluence for left-sided tumors. In the case of large tumors causing caudal compression of the kidney or stage III ACC with nodal involvement blocking the main lymphatic flow, an anterior lymphatic route to the interaortocaval nodes can also be activated as well as a cranial one, along the diaphragmatic vessels with possible direct flow to the thoracic duct and the mediastinum. The cranial limits of this "embryonal envelope" are represented by the diaphragmatic crux, while the caudal limits are represented by a transverse line from the caudal limit of the renal hilum to the aorta.

Some studies tracked the frequency of nodal metastatic involvement also in relation to different pathologic details such as disease stage [18, 19] and recognized that the right paracaval and left para-aortic nodes along with those at the renal hilum represent the first level of lymphatic drainage. In turn, these converge to the interaortocaval nodes which represent the second lymphatic level. Nodal involvement beyond the second level should be considered metastasis and the therapeutic strategy should be decided accordingly at multidisciplinary discussion. Interestingly, lymphatic metastases do not cross the aortic midline unless in case of very advanced stages or of previous excision of the interaortocaval nodes (recurrent disease). The involvement of anterior nodes such as those at the hepatic hilum is anecdotal. Bearing in mind these theoretical bases, it is possible to modulate the surgical strategy according to the clinical stage.

It seems wise to perform a "cavity" adrenalectomy for stage I ACC. This implies the en bloc removal of the adrenal gland with the cranial part of the renal capsule (anterior and posterior) and the retroperitoneal connective tissues. The caudal limit of this dissection is represented by the renal vein, the cranial limit by the diaphragm and the posterior one by the muscular plane. Lymphadenectomy includes the right posterior paracaval or the left para-aortic nodes along with those at the renal hilum, including the nodes positioned caudally and posterior to the renal vein and artery. When the tumor is small, the anterior dissection plane is generally easy to achieve maintaining a safety margin, and dissection from the liver, spleen and pancreas is not demanding. It is possible to perform this procedure with minimally invasive techniques, but the nodal dissection is easier with the robotic approach. Considering the confluence of the cited lymphatic structures into the main longitudinal lymphatic axis, attention should be paid to ligate or clip all longitudinal lymphatic structures.

For stage II–IV ACC we suggest a "regional" adrenalectomy. This consists of the removal of the adrenal gland *en bloc* with the entire renal capsule, of the lymph nodes at the renal hilum and of all retroperitoneal soft tissues surrounding the vena cava, including the aortocaval space which is dissected up to the aortic midline, from the diaphragm down to the inferior mesenteric artery, with exposure of the right aspect of the celiac trunk and superior mesenteric artery (right adrenalectomy, Figs. 9.1, 9.2, and 9.3). In the management of left-sided ACC, the medial limit of soft tissue clearance is the aortic midline, with exposure of the left side of visceral arteries; the clearance of interaortocaval lymphatic tissues is limited to those surrounding the left renal vein (Fig. 9.4).

Fig. 9.1 View of surgical field after right regional adrenalectomy

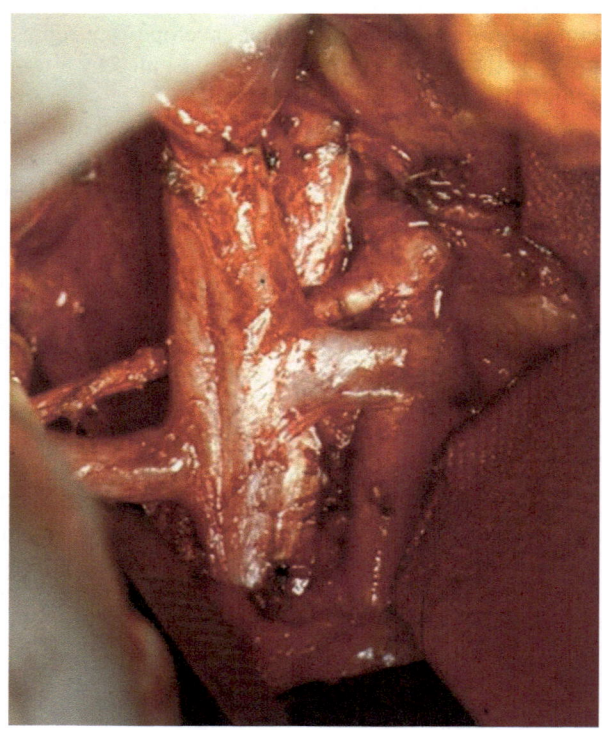

Fig. 9.2 Same patient of Fig. 9.1, detail of renal hilum after regional adrenalectomy

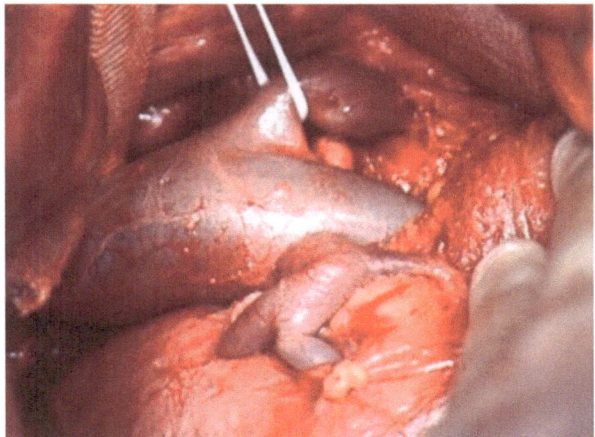

Fig. 9.3 Same patient of Figs. 9.1 and 9.2, detail of renal vessels after regional adrenalectomy

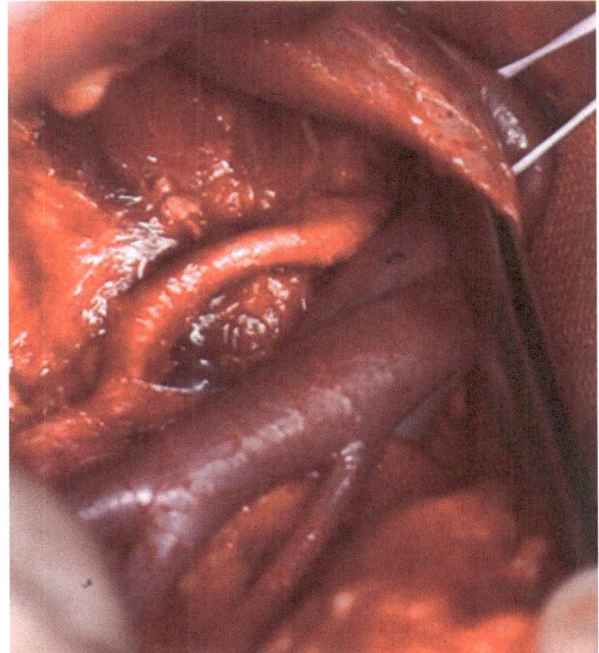

Fig. 9.4 View of surgical field after left regional adrenalectomy

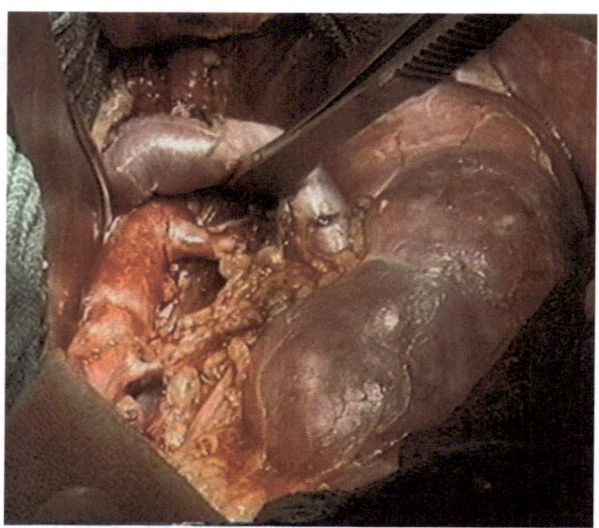

A large medialization of the liver or of the spleen-pancreas block is always the preliminary part of the procedure. Generally, it is advisable to perform the posterior dissection before dissection from the liver, spleen or pancreas, which is more easily performed once the entire block is better exposed. In the case of right-sided tumors we discourage dissection of the adrenal mass from the liver in favor of a resection of the liver, pursuing at least a sub-glissonian plane in the quest for radicality. The same is impossible for left-sided tumors, in which case dissection of the spleen and pancreas along with the splenic vessels should be meticulous to preserve tumor and margin integrity. Owing to their complexity, we perform these procedures in open surgery, through a subcostal incision. However, medialization of the spleen and pancreas may be extremely difficult in the case of large left-sided tumors as the spleen is often in a posterior position due to the tumor growth subtending the pancreas. In these conditions, it is wise to start with a laparoscopic approach with the patient in a right lateral decubitus: in this position the phrenosplenic ligament lays in an anterior position and is easily and completely dissected; the patient is then turned into a supine position and the procedure converted to open: the spleen will be free to follow all subsequent manipulations without any risk of capsule laceration. Resection of the surrounding organs such as the kidney, spleen or pancreas is only performed in the presence of direct infiltration.

9.3 Surgery for Recurrent Disease: The Guidelines

Unfortunately, recurrence is a common finding after surgery for ACC. It can present in different patterns: in the form of metastatic hematogenous disease (see also Chaps. 15 and 16) and in the form of tumor-bed or peritoneal recurrence.

On this subject the ESE-ENSAT guidelines [1] state the following:

R.8.5. In patients with recurrent disease and a disease-free interval of at least 12 months, in whom a complete resection/ablation seems feasible, we recommend surgery or alternatively other local therapies.

9.3.1 The Literature

Recurrence is frequent after surgery for ACC: a recent paper from the University of Michigan [20] reported a 70% recurrence rate in a series of 354 patients operated on for stage I–III ACC; recurrence was observed after a median disease-free survival (DFS) of 11 months. Surgery plays a pivotal role in the management of recurrence. In 1999, Schulick and Brennan, reporting the experience of the Memorial Sloan Kettering Cancer Center, focused on a group of 47 patients re-operated on (some of them multiple times, for a total of 83 re-resections). They showed that curative surgical treatment of recurrent ACC, both in the form of metastasectomy (lung, liver etc.) and/or ablation of tumor-bed recurrence and peritoneal metastases had a dramatic impact on survival: median survival was 74 months for those undergoing a R0 resection versus 16 months after incomplete resection; they also reported a 30-day mortality rate of 3.6% [21].

In the study of Glenn et al. [20], tumor-bed and peritoneal recurrence was detected in 28% and 20% of cases, as single-site recurrence or in the context of multiple-site recurrence in 11% and 17% of cases and in 6% and 14%, respectively; a metastatic pattern to parenchymal filters was observed in 52% of cases. Laparoscopic adrenalectomy was a risk factor for peritoneal metastases, and postoperative radiotherapy showed a protective effect on tumor-bed recurrence. A non-surgical treatment was chosen for 142 patients; they more likely had multiple metastases in the same organ, multiple-site recurrence and a short DFS (median 8 months). One hundred patients were operated on for recurrence; in this case after a median DFS of 17 months. The second operation was deemed R0 in 80% of cases but 79 patients recurred again after a median DFS of 6 months; tumor-bed and peritoneal metastases were exposed to a higher risk of further recurrence which, in turn, was observed in the same site in 67% of cases. Despite this, surgical management of recurrence had a positive impact on survival, which was appreciable up to the third operation. The German ACC Study Group studied 154 recurrent patients and reached similar conclusions [22]. They identified DFS after adrenal surgery >12 months and curative resection of recurrence as the most important predictors of survival. Colleagues from the University of Turin (Italy) stratified 106 patients with ACC recurrence, as having a unique (35% of cases), multiple within a single organ (20.8%) or multiple-site (43%) recurrence. Locoregional treatments were used in 100%, 68% and in 26% of patients in the three groups, respectively; these included surgery (86%), radiotherapy or radiofrequency ablation (9%) or multiple treatments. They reported a disease-free status in 60% of treated cases and a subsequent DFS of 15 months. The best survival results were achieved in the case of single lesion subjected to locoregional treatments [23].

9.3.2   Reasoning to Establish a Sound Surgical Strategy

Surgery for recurrent disease requires an eclectic approach due to the different possible scenarios. Resection of hepatic and lung metastases follows the specific general rules of liver and lung surgery for metastatic disease and their integration with ablative treatments such as radiofrequency ablation and radiotherapy; although these treatments will not be discussed in this chapter, we consider that the adrenal surgeon should be proficient in hepatic surgery. If this is not the case, the surgical team should include a hepatic surgeon. In the same way, lung metastases are normally treated by thoracic surgeons, in a two-step procedure if multiple-site metastases (thoracic and abdominal) are deemed resectable with curative intent. Peritoneal metastases are managed following the rules of peritonectomy for peritoneal diffusion of other primaries such as ovary, colonic or gastric cancer. Sacrifice of solid organs (liver, kidney, spleen, left pancreas, uterus and adnexa) and bowel is performed as required; in our experience, the extent of peritonectomy varies according to the clinical presentation and the technical approach to adrenalectomy. In fact, peritoneal metastases after open surgery require less extensive peritonectomies than those observed after minimally invasive surgery. The Peritoneal Cancer Index (PCI) should be always assessed at the beginning of the procedure and the Completeness of Cytoreduction (CC) should be measured at the end. It is our practice to extend these measures to tumor-bed recurrences even though, in strict terms, these are not peritoneal metastases.

Concerning the management of tumor-bed recurrence, two possible scenarios exist: the recurrence is expression of tumor seeding following a correct oncological resection, or, alternatively, it is expression of tumor persistency due to incomplete primary resection, with or without tumor seeding. In the first case the surgical target is represented by all tumoral nodules detected at cross-sectional imaging; resection of adjacent infiltrated organs is mandatory. Dissection of recurrent tumor from adjacent organs carries an extremely high risk of recurrence and should be contemplated only if the required resection exposes the patient to a disproportionate risk (i.e., pancreatic resection in a high-risk, fragile patient). In any case the resection should achieve tumor-free margins. In the second case, the re-operation should clear all nodules detected at imaging and achieve the required regional soft-tissue and lymphatic clearance not performed at first instance. In any case, these procedures require the complete mobilization of the liver and of the spleen-pancreas block; this latter may be facilitated by the need to resect the spleen and/or the left pancreas due to tumor invasion. The lymphadenectomy may also be facilitated by resection of other involved organs such as the kidney (Fig. 9.5). In all referral institutions these procedures are conducted in open surgery, with preference for a median laparotomy if peritoneal metastases are suspected.

Fig. 9.5 View of surgical field after enlarged regional demolition including nephrectomy, left pancreatectomy, splenectomy, and left colectomy for recurrent adrenocortical carcinoma

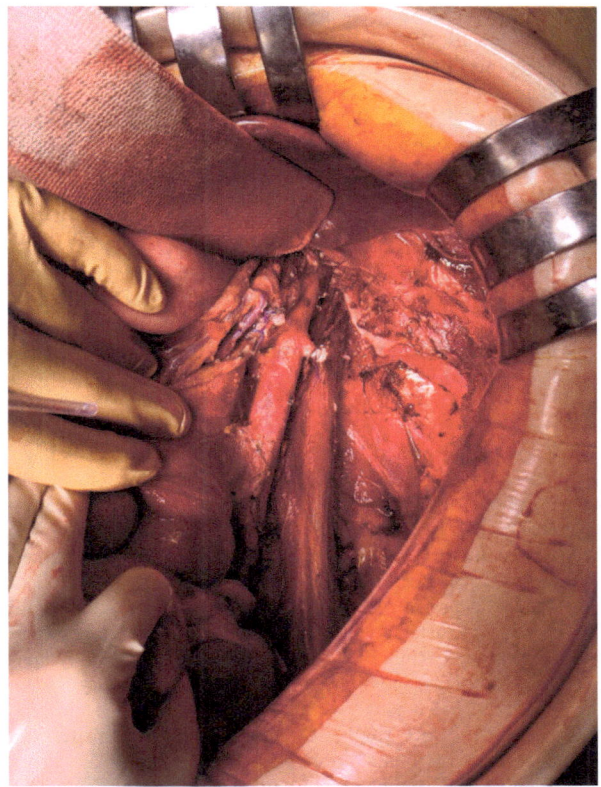

References

1. Fassnacht M, Dekkers OM, Else T, et al. European Society of Endocrinology Clinical Practice Guidelines on the management of adrenocortical carcinoma in adults, in collaboration with the European Network for the Study of Adrenal Tumors. Eur J Endocrinol. 2018;179(4):G1–G46.
2. Gaujoux S, Mihai R. European Society of Endocrine Surgeons (ESES) and European Network for the Study of Adrenal Tumours (ENSAT) recommendations for the surgical management of adrenocortical carcinoma. Br J Surg. 2017;104(4):358–76.
3. Fassnacht M, Assie G, Baudin E, et al. Adrenocortical carcinomas and malignant phaeochromocytomas: ESMO-EURACAN Clinical Practice Guidelines for diagnosis, treatment and follow-up. Ann Oncol. 2020;31(11):1476–90. Erratum in: Ann Oncol. 2023;34(7):631
4. Shah MH, Goldner WS, Benson AB, et al. Neuroendocrine and adrenal tumors, Version 2.2021. NCCN Clinical Practice Guidelines in Oncology. J Natl Compr Cancer Netw. 2021;19(7):839–68.
5. Caiazzo R, Marciniak C, Lenne X, et al. Adrenalectomy risk score: an original preoperative surgical scoring system to reduce mortality and morbidity after adrenalectomy. Ann Surg. 2019;270(5):813–9.
6. Gray WK, Day J, Briggs TWR, et al. Volume-outcome relationship for adrenalectomy: analysis of an administrative dataset for the Getting It Right First Time Programme. Br J Surg. 2021;108(9):1112–9.

7. Kerkhofs TM, Verhoeven RH, Bonjer HJ, et al. Surgery for adrenocortical carcinoma in The Netherlands: analysis of the national cancer registry data. Eur J Endocrinol. 2013;169(1):83–9.
8. Margonis GA, Kim Y, Prescott JD, et al. Adrenocortical carcinoma: impact of surgical margin status on long-term outcomes. Ann Surg Oncol. 2016;23(1):134–41.
9. Bilimoria KY, Shen WT, Elaraj D, et al. Adrenocortical carcinoma in the United States: treatment utilization and prognostic factors. Cancer. 2008;113(11):3130–6.
10. Crucitti F, Bellantone R, Ferrante A, et al. The Italian Registry for Adrenal Cortical Carcinoma: analysis of a multiinstitutional series of 129 patients. Surgery. 1996;119(2):161–70.
11. Panjwani S, Moore MD, Gray KD, et al. The impact of nodal dissection on staging in adrenocortical carcinoma. Ann Surg Oncol. 2017;24(12):3617–23.
12. Reibetanz J, Jurowich C, Erdogan I, et al. Impact of lymphadenectomy on the oncologic outcome of patients with adrenocortical carcinoma. Ann Surg. 2012;255(2):363–9.
13. Deschner BW, Stiles ZE, DeLozier OM, et al. Critical analysis of lymph node examination in patients undergoing curative-intent resection for adrenocortical carcinoma. J Surg Oncol. 2020;122(6):1152–62.
14. Gerry JM, Tran TB, Postlewait LM, et al. Lymphadenectomy for adrenocortical carcinoma: Is there a therapeutic benefit? Ann Surg Oncol. 2016;23:708–13.
15. Tseng J, DiPeri T, Chen Y, et al. Adrenocortical carcinoma: the value of lymphadenectomy. Ann Surg Oncol. 2022;29(3):1965–70.
16. Porpiglia F, Fiori C, Daffara FC, et al. Does nephrectomy during radical adrenalectomy for stage II adrenocortical cancer affect patient outcome? J Endocrinol Investig. 2016;39(4):465–71.
17. Bornstein SR, Allolio B, Arlt W, et al. Diagnosis and treatment of primary adrenal insufficiency: an Endocrine Society clinical practice guideline. Clin Endocrinol Metab. 2016;101(2):364–89.
18. Sada A, Glasgow AE, Lyden ML, et al. Informing therapeutic lymphadenectomy: Location of regional metastatic lymph nodes in adrenocortical carcinoma. Am J Surg. 2022;223(6):1042–5.
19. Reibetanz J, Rinn B, Kunz AS, et al. Patterns of lymph node recurrence in adrenocortical carcinoma: possible implications for primary surgical treatment. Ann Surg Oncol. 2019;26(2):531–8.
20. Glenn JA, Else T, Hughes DT, et al. Longitudinal patterns of recurrence in patients with adrenocortical carcinoma. Surgery. 2019;165(1):186–95.
21. Schulick RD, Brennan MF. Long-term survival after complete resection and repeat resection in patients with adrenocortical carcinoma. Ann Surg Oncol. 1999;6(8):719–26.
22. Erdogan I, Deutschbein T, Jurowich C, et al. The role of surgery in the management of recurrent adrenocortical carcinoma. J Clin Endocrinol Metab. 2013;98(1):181–91.
23. Calabrese A, Puglisi S, Borin C, et al. The management of postoperative disease recurrence in patients with adrenocortical carcinoma: a retrospective study in 106 patients. Eur J Endocrinol. 2023;188(1):118–24.

Adrenocortical Carcinoma with Vena Cava Involvement

10

Nazario Portolani, Franco Nodari, Guido A. M. Tiberio, and Stefano Bonardelli

Adrenocortical carcinoma (ACC) shows a significant propensity to invade vascular structures. Although vascular involvement may reveal itself in the form of direct infiltration of a vein, it more often consists of tumor growth inside the vascular lumen (15–25% of cases), forming a neoplastic thrombus [1]. Regardless of the presentation, venous involvement is a marker of aggressiveness which may bring into question the surgical indication.

10.1 The Guidelines

The guidelines issued in 2018 by the European Society of Endocrinology in collaboration with the European Network for the Study of Adrenal Tumors [2, 3] do not consider the presence of a tumor thrombus as a contraindication for surgery. They state: "We recommend that individualized treatment decisions are made in cases of tumors with extension into large vessels based on multidisciplinary surgical team. Such tumors should not be regarded "unresectable" until reviewed in an expert center". A main condition must, however, be respected: surgery must be performed by surgeons with special expertise in adrenal surgery for primary malignancy.

N. Portolani (✉) · G. A. M. Tiberio
General Surgery, Department of Clinical and Experimental Sciences, University of Brescia at ASST Spedali Civili di Brescia, Brescia, Italy
e-mail: nazario.portolani@unibs.it; guido.tiberio@unibs.it

F. Nodari
Vascular Surgery Unit, ASST Spedali Civili di Brescia, Brescia, Italy
e-mail: franco.nodari@spedalicivili.brescia.it

S. Bonardelli
Vascular Surgery, Department of Clinical and Experimental Sciences, University of Brescia at ASST Spedali Civili di Brescia, Brescia, Italy
e-mail: stefano.bonardelli@unibs.it

© The Author(s) 2025
G. A. M. Tiberio (ed.), *Primary Adrenal Malignancies*, Updates in Surgery,
https://doi.org/10.1007/978-3-031-62301-1_10

The ESMO-EURACAN guidelines issued in 2020 state: "In order to obtain an R0 resection of a locally advanced ACC, it may be necessary to resect (parts of) adjacent organs such as the wall of the vena cava, liver, spleen, colon, pancreas and/ or stomach. Complete en bloc resection of the tumoral mass, including periadrenal fat and adjacent organs if necessary, is mandatory to avoid tumor rupture or spillage that portends an adverse outcome" [4].

10.2   The Multidisciplinary Team

It is general opinion that patients with ACC benefit from a multidisciplinary management by a team of experts from the beginning. For surgical planning, the extension of the thrombus and the foreseeable technical solution determine the strategy and suggest the composition of the surgical team, which may include a vascular and/or a cardiac surgeon. Proven expertise in both adrenal and oncologic surgery, including major liver procedures, is needed in view of the specific anatomy, the malignant character of the disease and the potential need for multiorgan en bloc resection. This guarantees the best oncologic standard and the lowest risk of complications. The anesthesiologic team should be involved from the preliminary discussion of the surgical strategy. Dedicated anesthetists are required, with the experience and skills needed to face any complication arising from major splanchnic, vascular and cardiac procedures. Intraoperatively, an expert in transesophageal ultrasonography is required to monitor the stability of the endoluminal thrombus during surgery and to guarantee hemodynamic monitoring if the inferior vena cava (IVC) is clamped above the suprahepatic veins.

10.3   The Cancer and the Patient

Clinical signs of vena cava obstruction are rare, as is neoplastic pulmonary embolism; the flow in the IVC, albeit reduced, persists even if imaging suggests venous thrombosis, which is reported in about 5% of cases [3]. Imaging is characteristic: the neoplastic thrombus appears as an endoluminal growth with contrast medium enhancement at computed tomography (CT); it may fill and dilate the vessel lumen (Fig. 10.1). Its extension is graded as follows: Level I, the thrombus reaches the adrenal or renal vein (left side); Level II, the IVC is occupied up to the level of the hepatic veins; Level III, the thrombus reaches the suprahepatic IVC; Level IV, the right cardiac chamber is occupied by the tumor thrombus. Direct infiltration of the IVC should be suspected whenever the margins of the lesion and the vena cava are not clearly detectable. IVC involvement is typically observed in the presence of large (≥ 10 cm) and right-sided adrenal masses, as the length of the left renal vein has a protective role [5]. Retrograde non-neoplastic thrombosis is a common finding; it can involve the left renal vein up to the confluence of the gonadal vein.

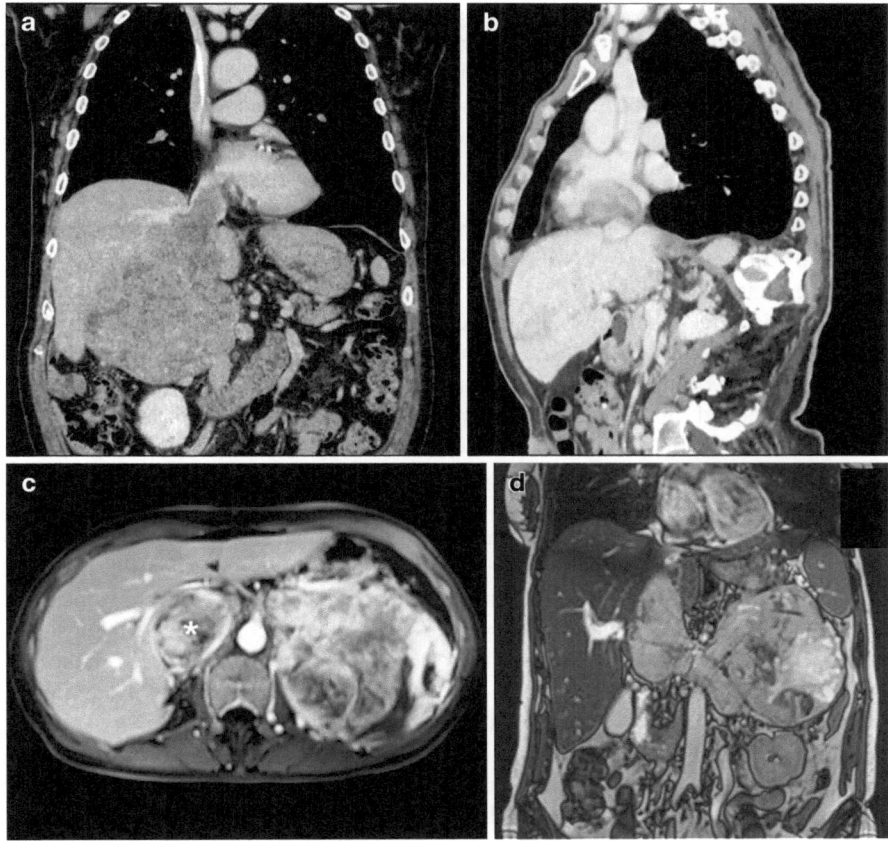

Fig. 10.1 Coronal (**a**) and sagittal (**b**) venous-phase computed tomography (CT) scan show a right adrenocortical carcinoma (ACC) with neoplastic thrombus in the inferior vena cava (IVC) extending to the right cardiac chamber (**b**). Axial 3D gradient-echo (GRE) fat-suppressed T1w after contrast administration (**c**) and coronal T2w (**d**) magnetic resonance imaging (MRI) sequences show an enlarged left renal vein and IVC (*asterisk*) due to the presence of inhomogeneous neoplastic tissue in its lumen originating from a left ACC. Note the intrinsically better resolution of contrast-enhanced MRI compared to CT

10.4 Technical Aspects

In surgical planning, two phases must be considered: the treatment of the tumor and the management of the vascular invasion.

Difficulties arise from the size and soft consistency of the tumor, the activation of collateral circulation, and the need to perform an en bloc resection of the involved tissues and organs. The literature does not support preoperative embolization to reduce bleeding [6]; some authors advocate early ligature of the adrenal artery to favor the collapse of the collateral circulation [7].

Open surgery is standard [1]. When surgery can be completed via an abdominal approach (the edge of the thrombus remains below the diaphragmatic plane), a bilateral subcostal incision is preferred, with median extension to the xiphoid if needed. When the edge of the thrombus reaches the pericardium, the indication for a wider approach must be discussed. Median sternotomy is mandatory when the thrombus reaches the right atrium, but this is not the rule in any other case: the IVC can be controlled within the pericardium through the diaphragm.

The first phase of the procedure pursues the mobilization of the adrenal gland. Care must be taken to avoid mobilization/fragmentation of the thrombus during the different phases of the procedure, such as liver mobilization and dissection of the suprahepatic veins, maneuvers along the IVC and, above all, positioning of the vascular clamp above the edge of the thrombus. This phase must be always monitored by intraoperative transesophageal ultrasound which may identify unstable or floating thrombus and thus suggest safety maneuvers such as an early clamping of the vein. In any case, the risk of thrombus fragmentation seems to be low and intraoperative neoplastic pulmonary embolism is uncommon [1, 4]. The ipsilateral kidney should be preserved if not infiltrated.

The liver is often enlarged and congested, and pushed toward the diaphragm by the tumor. The advice is not to pursue the dissection plane between the liver and adrenal but to resect the liver in order to optimize radicality. The exact extension of IVC involvement can be verified only after complete dissection of its anterior surface, which is approached with the "liver transplantation technique" [8]. However, differing from liver surgery, any maneuver can be more difficult than usual, in particular if the IVC is dilated and filled by the tumor, and the small accessory veins are particularly short. Nonetheless, the IVC must be completely dissected well above the edge of the thrombus, if necessary above the three hepatic veins, to facilitate positioning of the clamp and allow thrombectomy of the hepatic veins if required. The vascular dissection above the liver is generally easy as the vena cava at this level is rarely infiltrated. In the case of fusion of the lateral or posterior surface of the IVC with the tumor, an anterior approach to the IVC with a subsequent right hepatectomy has been proposed [9, 10]. This technique does not exclude the need to dissect at least the anterior aspect of the IVC; furthermore, such a major liver resection increases the risk of operative morbidity and is not justified from the oncological point of view, considering that microscopic liver involvement occurs in 40% of cases [10]. In our opinion, this approach must be strictly reserved for selected patients. Once dissected free from the kidney, liver and posterior muscular plane en bloc with the perirenal capsule, the adrenal mass remains attached to the IVC, which is dissected free, mobilized and taped above and below the involved tract. Attention must be focused on the accurate control of the lumbar veins, which are hypertrophic if a collateral circulation has been activated.

The "vascular" phase requires an accurate strategy, according to the type and extension of vascular involvement reported by preoperative cross-sectional imaging.

In cases of parietal infiltration, the strategy is dictated by its extension. A lateral vascular clamp, after complete posterior mobilization, allows for preservation of caval flow, for a marginal vein resection and for an easy reconstruction, by direct suture or with patch interposition, of limited infiltrations around or near the adrenal vein confluence. In the event of major infiltration of the IVC, its prosthetic replacement is preferable for oncologic reasons.

When a thrombus is present, the strategy varies according to its cranial extension. Thrombectomy is an "at risk of bleeding" maneuver and the anesthesiologic team should be ready for intraoperative blood transfusion. The use of intraoperative blood recovery is debated but, considering fragmentation of the neoplastic thrombus as a systematic occurrence, it should be reserved for very particular cases.

Correct clamp positioning is critical. If the ostia of the renal veins are spared, clamp positioning above their confluence allows for a retrograde flow which preserves the renal parenchyma. When facing a left-sided cancer, the caudal vascular clamp should be positioned below the left renal vein in an oblique position, so as to preserve the retrograde venous flow of the right kidney. If this is not possible or if the thrombus involves the right renal vein, after arterial exclusion of the right kidney, complete clearing of the ostium of the renal veins may be accomplished from inside the IVC, clamped below the right renal vein. In this case, the IVC reconstruction will begin at the lower edge of the cavotomy and the clamp is promptly repositioned above the renal veins allowing reperfusion of the kidney. Thrombectomy of the IVC and ostium of the left renal vein allows the positioning of a clamp at the renal vein confluence and subsequent thrombectomy of the renal vein. As an alternative, if prosthetic renal vein replacement is considered excessive, it is possible to resect the left renal vein provided that the gonadal vein, generally enlarged, can remain in situ, allowing for renal flow diversion toward the iliac vein. Furthermore, if the vena cava is infiltrated in proximity to the renal confluence, the renal veins can be resected and anastomosed to the prosthetic vena cava replacement (Fig. 10.2).

Rapid clamp repositioning has also a crucial role when the cranial clamp is placed above the suprahepatic veins: after clearing at least the cranial part of the thrombus and, if indicated, the ostium of the hepatic veins, the suture of the caval

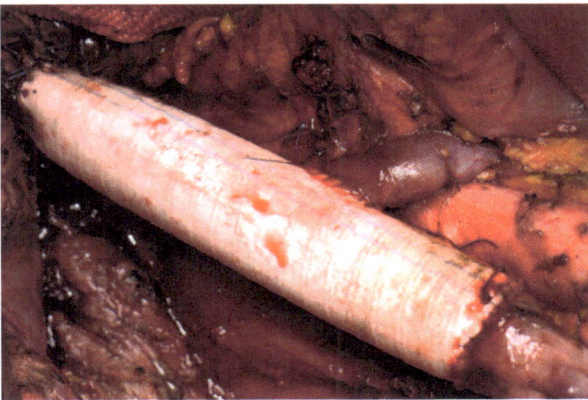

Fig. 10.2 Polytetrafluoroethylene (PTFE) graft replacement of the inferior vena cava. The left renal vein was anastomosed to the prosthetic graft

wall starts from the upper edge of the cavotomy and, as soon as it passes the supra-hepatic veins caudally, the clamp is repositioned below them, allowing restoration of splanchnic flow. In fact, if exclusion of the IVC flow is generally well tolerated, its association with the sudden drop of splanchnic flow must be discussed during the multidisciplinary planning meeting to implement any required compensatory measures. At this point hemodynamic monitoring by transesophageal ultrasound plays a critical role for the anesthetist, in regulating volume needs and amino support. If hemodynamic failure develops (some minutes of compensation are needed) a veno-venous bypass may be required. This must be prepared in advance: arranging the venous accesses during the preliminary phases of the operation allows for rapid activation of the bypass when required. Some authors advocate the "milking" maneuver (i.e., manual dislodgment of the edge of the thrombus downward) with the aim of directly applying the clamp below the suprahepatic veins or accelerating clamp repositioning [7, 10]. This maneuver may expose to thrombus fragmentation and dislodgment in the suprahepatic veins if incorrectly performed.

Management of level III and IV thrombus is more complex. Circulatory arrest is needed if the cardiac cavities must be approached and an aortopulmonary bypass must be activated. Atriotomy makes it possible to clear the cardiac chamber as well as the highest portion of the IVC and to easily control the hepatic vein confluence from above (Fig. 10.3), facilitating clamp positioning below the hepatic veins. The advantage of operating in a blood-free field without time-related constraints is counterbalanced by the heparinization and consequent derangement of coagulation, in particular if a hypothermic strategy is implemented. In these cases, a meticulous hemostasis must be pursued on all dissection planes before pump activation. A limited extension of the thrombus inside the right atrium may be controlled by a clamp inserted tangentially just above the thrombus, thus avoiding cardiac arrest; in any case, in these conditions collaboration with the cardiac surgeon is mandatory.

Cavotomy is regulated by the thrombus extension. Generally, it is possible to spare the IVC itself; sometimes a small resection of the vein's wall including the

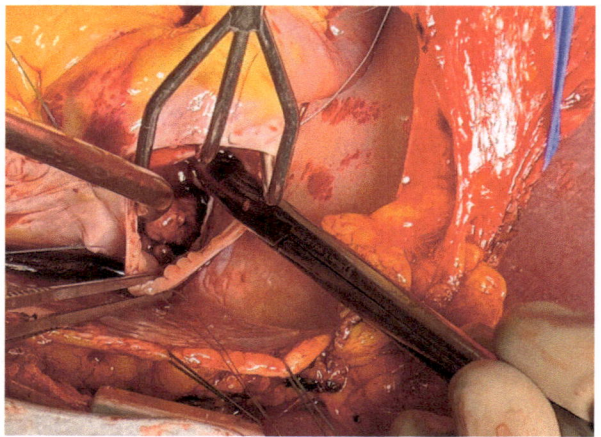

Fig. 10.3 View of the suprahepatic inferior vena cava through a right atriotomy, after removal of the atrial edge of the neoplastic thrombus. The caval thrombus is still in place

confluence of an apparently infiltrated right adrenal vein must be performed. In the presence of massive IVC infiltration there are no alternatives to its resection.

Reconstruction is the final phase of the procedure, as restoration of caval flow must always be pursued. Even though IVC ligation may be tolerated, especially after sacrifice of the right kidney, it has a major impact on quality of life; for this reason, it must be considered an extreme option. Under the protection of extracorporeal circulation or after splanchnic flow restoration, the reconstruction—or its completion—must be done paying special attention to its technical quality, without time constraints. Direct reconstruction of the IVC is generally possible without stenosis, especially if the vein was dilated by the intraluminal thrombus, but stenoses up to 50% of the lumen are tolerated (Fig. 10.4). The positioning of prosthetic patches or IVC replacement should be chosen according to the different clinical presentations. Bovine pericardium is generally preferred for prosthetic patches; the same material can be used for replacement of the vena cava, especially in the case of major discrepancies between the diameter of the two stumps, as an alternative to the widely used polytetrafluoroethylene (PTFE) graft or human homograft (Fig. 10.5).

Fig. 10.4 Direct reconstruction of the inferior vena cava after its resection. A stenosis <50% is appreciable. The postoperative course was uneventful

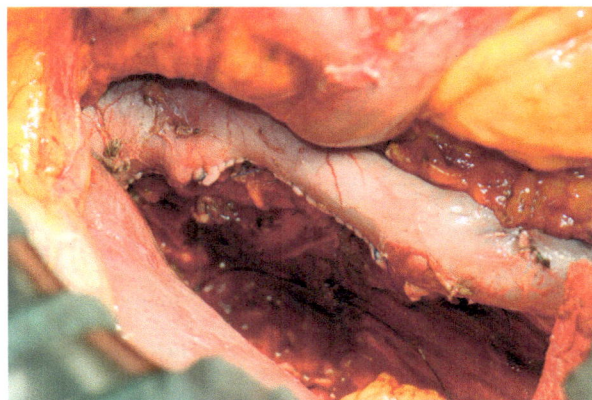

Fig. 10.5 Homograft replacement of the inferior vena cava

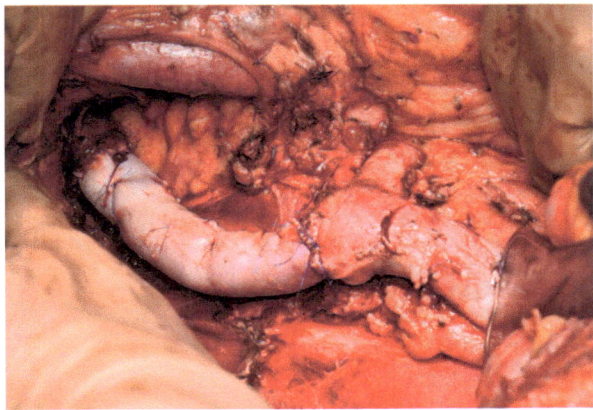

10.5 Results

Operative mortality ranges between 4% and 13% and is linked to hemorrhagic complications [1, 5], which represent the only cause of death in some experiences [2]. Long-term survival may be obtained after radical surgery. In the collective review by Chiche, 35% of patients survived at least 24 months [1]; in a subgroup analysis including only locally advanced ACC patients with complete (R0) resection, Laan et al. [11] reported similar survival at 24 months between patients with vena cava tumor thrombus and non-vena cava involvement, but at 36 months survival was better for the group with non-vena cava involvement (65% vs. 29%).

10.6 Final Considerations

Complete resection of the tumor remains the standard of care for adrenal cancer. The presence of IVC involvement is a further complication for the surgical procedure, but a technical solution is generally possible. The multidisciplinary evaluation, under the direction of the oncologist, has the role of defining the right indication, considering both the risk of surgery and the poor prognosis. A further evaluation of the medical regimen in the pre/postoperative setting, probably based on mutation pattern and gene expression, will be able to further select the patients for this complex surgery.

References

1. Chiche L, Dousset B, Kieffer E, Chapuis Y. Adrenocortical carcinoma extending into the inferior vena cava: presentation of a 15-patient series and review of the literature. Surgery. 2006;139(1):15–27.
2. Fassnacht M, Dekkers OM, Else T, et al. European Society of Endocrinology Clinical Practice Guidelines on the management of adrenocortical carcinoma in adults, in collaboration with the European Network for the Study of Adrenal Tumors. Eur J Endocrinol. 2018;179(4):G1–G46.
3. Gaujoux S, Mihai R. European Society of Endocrine Surgeons (ESES) and European Network for the Study of Adrenal Tumours (ENSAT) recommendations for the surgical management of adrenocortical carcinoma. Br J Surg. 2017;104(4):358–76.
4. Fassnacht M, Assie G, Baudin E, et al. Adrenocortical carcinomas and malignant phaeochromocytomas: ESMO-EURACAN Clinical Practice Guidelines for diagnosis, treatment and follow-up. Ann Oncol. 2020;31(11):1476–90. Erratum in: Ann Oncol. 2023;34(7):631
5. Turbendian HK, Strong VE, Hsu M, et al. Adrenocortical carcinoma: the influence of large vessel extension. Surgery. 2010;148(6):1057–64.
6. Ekici S, Ciancio G. Surgical management of large adrenal masses with or without thrombus extending into the inferior vena cava. J Urol. 2004;172(6 Pt 1):2340–3.
7. Ciancio G, Farag A, Gonzalez J, Gaynor JJ. Adrenal tumors of different types with or without tumor thrombus invading the inferior vena cava: an evaluation of 33 cases. Surg Oncol. 2021;37:101544.
8. Delis SG, Bakogiannis A, Ciancio G, Soloway M. Surgical management of large adrenal tumours: the University of Miami experience using liver transplantation techniques. BJU Int. 2008;102(10):1394–9.

9. Donadon M, Abdalla EK, Vauthey JN. Liver hanging maneuver for large or recurrent right upper quadrant tumors. J Am Coll Surg. 2007;204(2):329–33.
10. Coppa J, Citterio D, Cotsoglou C, et al. Transhepatic anterior approach to the inferior vena cava in large retroperitoneal tumors resected en bloc with the right liver lobe. Surgery. 2013;154(5):1061–8.
11. Laan DV, Thiels CA, Glasgow A, et al. Adrenocortical carcinoma with inferior vena cava tumor thrombus. Surgery. 2017;161(1):240–8.

Adrenocortical Carcinoma: The Posterior Minimally Invasive Approach

11

Pier Francesco Alesina, Polina Knyazeva, and Martin K. Walz

11.1 Introduction

Application of minimally invasive surgery in the surgical management of adreno-cortical carcinoma (ACC) sparks ongoing debate, despite some evidence indicating its potential to achieve comparable outcomes to open surgery, particularly among patients falling within stages I and II according to the European Network for the Study of Adrenal Tumors (ENSAT) [1, 2]. Nevertheless, it is important to acknowledge that the quality of evidence derived from these observational studies is regarded as very limited [3]. The specific role of the posterior retroperitoneoscopic adrenal-ectomy (PRA) remains unclear, given the scarcity of literature detailing its outcomes in the context of ACC treatment. In this chapter we describe the potential indications and the surgical technique for PRA in the treatment of ACC.

11.2 Surgical Technique

The description of PRA, which we pioneered, was first documented in 1995 [4], with subsequent significant refinements outlined in 2006 [5]. The patient is positioned in a prone, half-jack-knife position, with the lower legs forming a 90° angle with the thighs. To provide the necessary access below the ribcage, a rectangular cushion is inserted between the operating table and the patient's abdomen, allowing

P. F. Alesina (✉)
Universität Witten-Herdecke, Department of Endocrine Surgery, Helios Universitätsklinikum Wuppertal, Wuppertal, Germany
e-mail: pier.alesina@uni-wh.de

P. Knyazeva · M. K. Walz
Department of Surgery and Centre of Minimally Invasive Surgery, Evang. Kliniken Essen-Mitte, Essen, Germany
e-mail: polinakn@gmail.com; mk.walz@kem-med.com

G. A. M. Tiberio (ed.), *Primary Adrenal Malignancies*, Updates in Surgery,
https://doi.org/10.1007/978-3-031-62301-1_11

the abdominal wall to dangle anteriorly. Alternatively, a bolster can be placed beneath the chest and pelvis. A skin incision of about 2 cm in length is made at the level of the 12th rib, and access to the retroperitoneal space is achieved through a combination of blunt and sharp dissection using scissors. The retroperitoneum is further opened, and the space expanded by introducing a finger. Subsequently, a 5-mm port is carefully inserted just beneath the tip of the 11th rib under digital guidance. A blunt trocar equipped with an inflatable balloon and an adjustable sleeve (Medtronic, Minneapolis, USA) is introduced into the initial incision and securely positioned. Carbon dioxide insufflation is initiated, beginning with a pressure of 20 mmHg, which may be adjusted up to 30 mmHg depending on the patient's characteristics, such as their degree of obesity and the extent of retroperitoneal fatty tissue, as well as the size of the tumor. The creation of a working space involves the dissection of Gerota's fascia and the gentle displacement of retroperitoneal fatty tissue in a ventral and caudal direction. This maneuver can already allow for the visualization of both the kidney and the adrenal gland. A third trocar (5- or 10-mm) is inserted medially, positioned below the 12th rib, all while maintaining visual control and taking care to avoid any damage to the subcostal nerve that runs parallel to the rib. Therefore, the tip of the third port should be blunt. The dissection commences on the lateral aspect of the upper pole of the kidney, with the objective of mobilizing the kidney and releasing it from any adhesions to the retroperitoneal fatty tissue. Subsequently, the kidney is retracted and turned in a caudal and medial direction to expose the lower pole of the adrenal gland. Depending on the distinct anatomical position of the left and right adrenal glands, the mobilization of the kidney should be more extensive on the left side. Extremely helpful are modern bipolar or ultrasonic instruments. The dissection proceeds from lateral to medial taking down the fatty tissue between kidney and adrenal gland. By this horizontal dissection on the anterior Gerota's fascia feeding branches from the renal artery are cut. On the right side the inferior vena cava (IVC) is identified medially, usually close to the entrance of the right renal vein, which is directly visualized only in cases of very large or caudally located adrenal tumors. Afterward the vertical retrocaval dissection is meticulously carried out until reaching the adrenal vein. It is common to encounter small arteries running horizontally behind the IVC, that are all typically dissected and divided. On the left side, the adrenal vein is isolated medially to the lower pole of the adrenal gland. In cases where adrenalectomy is performed due to suspected ACC, a modification of the technique may be considered to extend the resection in order to include regional lymph nodes. On the right side, the renal artery and vein can be identified, and the dissection line closely follows the renal vessels until reaching the inter-aortocaval region, enabling the removal of fatty tissue that encompasses the para-adrenal and inter-aortocaval lymph nodes. This meticulous dissection exposes the entire posterior wall of the vena cava, facilitating the continuation of lymphadenectomy until the right crux is reached. Approximately at the upper third of the adrenal gland, the short adrenal vein is then isolated and divided. On the left side, the renal artery is identified and dissected free until its origin at the aorta is reached. The removal of the left paraaortic fatty tissue, along with the lymph nodes, is carried out, allowing for visualization of the aorta up to the level of the left crux. During this step, the left adrenal vein, which converges into the renal vein, and the inferior diaphragmatic vein are naturally identified. The division of the adrenal vein is routinely performed using energy devices. Subsequently, the cranial dissection marks the next stage of the procedure, and it is followed by dissection of the posterior peritoneal layer on the right side and the lamina between the pancreas and the adrenal gland on the left side. These layers may be excised en bloc with the tumor, revealing the right posterior liver segments on the right side and the pancreas along with the splenic vessels on the left side. The opening of the peritoneal layer does not significantly impede the operating space, ensuring the safe completion of the procedure. The specimen is carefully placed within a retrieval bag and pulled through the middle incision. If necessary, in-bag fragmentation of the specimen can be performed to facilitate its extraction. Suction drains are typically not utilized.

11.3 Discussion

The use of minimally invasive techniques in the surgical treatment of ACC remains a topic of ongoing discussion. Current guidelines advocate open surgery as the standard approach for confirmed or highly suspected ACC. However, for tumors measuring less than 6 cm without any sign of local invasion, minimally invasive adrenalectomy, adhering to oncological principles, can be considered [6]. While there is no conclusive evidence in the literature, experts generally believe that a minimum annual caseload of six adrenalectomies per year is necessary to maintain sufficient proficiency in adrenal surgery. However, for those engaged in ACC surgery, performing over 20 adrenalectomies per year is highly desirable [6]. An analysis of data from the US National Cancer Database from 2010 to 2014 revealed that among 196 patients who underwent attempted laparoscopic adrenalectomy for ACC, 38 individuals (19.4%) required conversion to an open surgical procedure [7]. Notably, risk factors for conversion included larger tumor size (9 cm) and the need for a right-sided adrenalectomy. The conversion rate was lower in patients who underwent robotic procedures (10.9%) compared to those who had laparoscopic procedures (22.0%), with a p-value of 0.095. This suggests that the utilization of articulated arms in robotic surgery may aid in dissection, particularly in cases involving larger tumors, potentially reducing the need for extensive manipulation. Furthermore, the same study demonstrated that conversion significantly impacted overall survival at 2 years and was associated with a higher incidence of R1 resections. No studies in the literature have reported the outcomes of PRA for ACC in a cohort of patients. Previous studies from our group included single cases treated by PRA and do not allow any conclusion about the oncological appropriateness of the procedure. In a paper from 2005, the experience with 33 PRA for primary adrenal tumors >6 cm selected from a prospective series of 429 minimally invasive adrenal operations was published [8]. There were no perioperative deaths. Patients with large tumors had an increased conversion rate ($p = 0.039$), longer operating time ($p < 0.001$) and greater intraoperative blood loss ($p = 0.007$) than those with smaller lesions, but a similar overall morbidity rate (p = 0.207). Six malignant tumors were identified in this series (diameter 4–10 cm; four pheochromocytomas and two ACCs). Local recurrence developed in two patients and distant metastasis occurred in all six patients with malignant lesions. Given that minimally invasive adrenalectomy has showcased numerous advantages over open surgery in the treatment of the majority of adrenal masses, including reduced blood loss, shorter recovery periods, and diminished postoperative pain [9], it seems reasonable to consider this approach for localized (stage I–II) ACC. As per several studies [10, 11], which found no significant differences in oncological outcomes, it can be concluded that laparoscopic adrenalectomy may be considered a viable choice for treating localized ACC, provided that appropriate patient selection criteria are met and surgical adherence to oncological principles is maintained. However, it is essential to evaluate these findings with caution due to the retrospective nature of the studies and the relatively small number of patients included. The maximum tumor diameter that can be safely approached using minimally invasive techniques remains a subject of uncertainty. For laparoscopic adrenalectomy, a proposed limit of 10 cm has been suggested [10]. This size limit appears to be acceptable also for PRA. However, it is challenging to provide a definitive recommendation due to the absence of published data in the literature. Therefore, any opinion on this matter can only be offered based on the experience gathered over the years. Opting for a minimally invasive approach when dealing with a tumor larger than 6 cm should be done cautiously, taking into account several individual factors. These factors include the patient's dimensions, the quantity of retroperitoneal fatty tissue, the potential presence of enlarged lymph nodes, and the surgeon's personal experience and proficiency with the technique (Fig. 11.1). This is especially critical when conducting minimally invasive lymphadenectomy or multivisceral resection. PRA allows for the feasible execution of left para-aortic and right paracaval lymphadenectomy, extending to the hilum of the kidney and the infrarenal region. This technique offers a direct path to the retroperitoneal area, bypassing the need to mobilize intra-abdominal organs and reducing the risk of potential injuries of intra-abdominal organs (Fig. 11.2). However, it is crucial to

Fig. 11.1 Computed tomography scan of right-sided adrenocortical carcinoma, 35 W. Posterior retroperitoneoscopic approach contraindicated because safe circumferential dissection impossible due to tumor size (15 cm ∅)

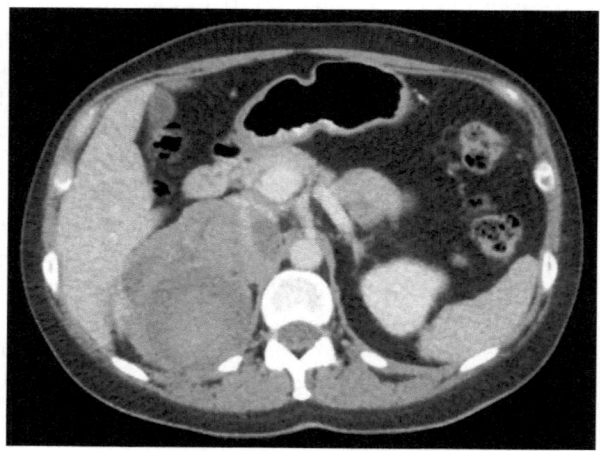

Fig. 11.2 Computed tomography scan of left-sided adrenocortical carcinoma, 36 M. En bloc resection including regional lymphadenectomy by posterior retroperitoneoscopic approach (*red dotted line*)

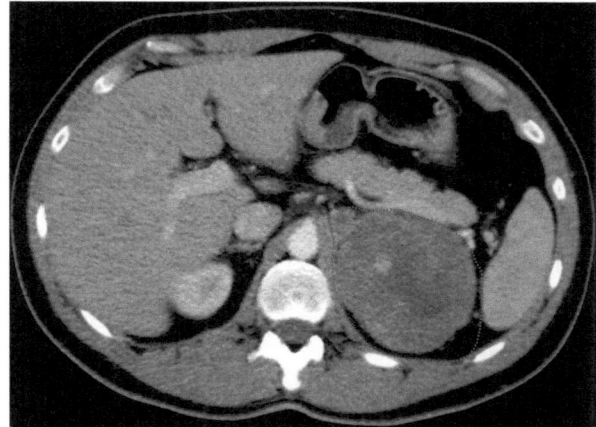

Fig. 11.3 Magnetic resonance imaging of right-sided adrenocortical carcinoma with tumor thrombus in inferior vena cava (*arrow*), 20 W. Posterior retroperitoneoscopic adrenalectomy including removal of tumor thrombus by adrenal vein excision

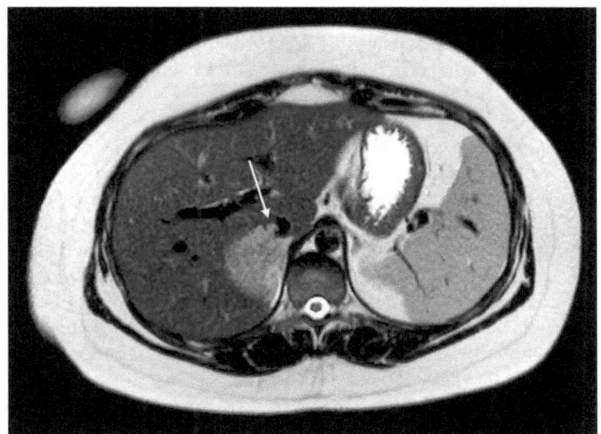

note that in the event of suspected involvement of interaortocaval lymph nodes in the case of left-sided tumors or contralateral lymph nodes (for both sides), a bilateral retroperitoneoscopic approach should be taken into account allowing a systematic lymphadenectomy. Patients whose preoperative imaging strongly suggests infiltration of adjacent organs, necessitating the consideration of multiorgan or compartment resection, are typically not suitable candidates for minimally invasive approaches as well. Nevertheless, infiltration of adjacent organs, primarily the kidney, is often only suspected at preoperative imaging and an endoscopic approach is still possible in many cases. Early conversion to open surgery should be considered in cases of intraoperative confirmation of infiltration. In ACCs with small tumor thrombus in the IVC, we demonstrated a complete venous vascular control by PRA (Fig. 11.3). By clamping the IVC segment between conjunction of the renal veins and high retrohepatic, an en bloc resection of ACC and thrombus can be performed by excision of the adrenal vein at entering the IVC. The venous defect has to be closed with a running suture [12]. Of course, this approach is reserved for highly selected patients and requires advanced skills in endoscopic surgery.

11.4 Conclusions

PRA proves to be both feasible and safe for patients diagnosed with adrenal pathologies and demonstrated to have a similar complication rate in patients with larger tumors. It can be considered a viable alternative to open or laparoscopic adrenalectomy, especially for patients with tumors measuring 10 cm or less. This technique also allows for lymphadenectomy and the resection of more advanced tumors, but only in carefully selected patients. Nevertheless, long-term oncological results are still missing.

References

1. Brix D, Allolio B, Fenske W, et al. Laparoscopic versus open adrenalectomy for adrenocortical carcinoma: surgical and oncologic outcome in 152 patients. Eur Urol. 2010;58(4):609–15.
2. Porpiglia F, Fiori C, Daffara F, et al. Retrospective evaluation of the outcome of open versus laparoscopic adrenalectomy for stage I and II adrenocortical cancer. Eur Urol. 2010;57(5):873–8.
3. Nakanishi H, Miangul S, Wang R, et al. Open versus laparoscopic surgery in the management of adrenocortical carcinoma: a systematic review and meta-analysis. Ann Surg Oncol. 2023;30(2):994–1005.
4. Walz MK, Peitgen K, Krause U, Eigler FW. Die dorsale retroperitoneoskopische Adrenalektomie—eine neue operative Technik [Dorsal retroperitoneoscopic adrenalectomy—a new surgical technique]. Zentralbl Chir. 1995;120(1):53–8.
5. Walz MK, Alesina PF, Wenger FA, et al. Posterior retroperitoneoscopic adrenalectomy--results of 560 procedures in 520 patients. Surgery. 2006;140(6):943–50.
6. Fassnacht M, Dekkers OM, Else T, et al. European Society of Endocrinology Clinical Practice Guidelines on the management of adrenocortical carcinoma in adults, in collaboration with the European Network for the Study of Adrenal Tumors. Eur J Endocrinol. 2018;179(4):G1–G46.
7. Delozier OM, Stiles ZE, Deschner BW, et al. Implications of conversion during attempted minimally invasive adrenalectomy for adrenocortical carcinoma. Ann Surg Oncol. 2021;28(1):492–501.
8. Walz MK, Petersenn S, Koch JA, et al. Endoscopic treatment of large primary adrenal tumours. Br J Surg. 2005;92(6):719–23.
9. Brunt LM, Doherty GM, Norton JA, et al. Laparoscopic adrenalectomy compared to open adrenalectomy for benign adrenal neoplasms. J Am Coll Surg. 1996;183(1):1–10.
10. Donatini G, Caiazzo R, Do Cao C, et al. Long-term survival after adrenalectomy for stage I/II adrenocortical carcinoma (ACC): a retrospective comparative cohort study of laparoscopic versus open approach. Ann Surg Oncol. 2014;21(1):284–91.
11. Gaillard M, Razafinimanana M, Challine A, et al. Laparoscopic or open adrenalectomy for stage I-II adrenocortical carcinoma: a retrospective study. J Clin Med. 2023;12(11):3698.
12. Walz MK, Jongekkasit I, Alesina PF. Retroperitoneoscopic adrenalectomy in adrenocortical carcinoma with tumor thrombus in inferior vena cava: one step further in minimally invasive endocrine surgery. Video Endocrinol. 2021;8(1) https://doi.org/10.1089/ve.2020.0202.

Surgery for Malignant Pheochromocytoma

12

Giovanni Casole, Silvia Ministrini, Federica Gabella, and Guido A. M. Tiberio

12.1 The Guidelines

According to the 2020 European Society for Medical Oncology guidelines [1], 2019 Associazione Italiana di Oncologia Medica guidelines [2] and 2014 Endocrine Society guidelines [3], surgical excision is the first line treatment for pheochromocytomas (PHEOs).

12.2 Indication for Surgery

Preoperative diagnosis of malignant PHEO in patients without metastatic disease or clear evidence of locoregional invasion is challenging, even though it does not impact the therapeutic strategy. Surgical excision is recommended in all PHEOs due to the catecholamine-related risks.

Regardless of the type of approach and preoperative diagnosis, proper resection must involve complete excision of the neoplasm without spillage of the mass. It is important to minimize manipulation to reduce hypertensive spikes [4].

The main outcome of complete surgical resection (R0) is to cure or increase overall survival.

Supplementary Information The online version contains supplementary material available at https://doi.org/10.1007/978-3-031-62301-1_12.

G. Casole (✉) · S. Ministrini · F. Gabella · G. A. M. Tiberio
General Surgery, Department of Clinical and Experimental Sciences, University of Brescia at ASST Spedali Civili di Brescia, Brescia, Italy
e-mail: giovannicasole@gmail.it; ministrini.silvia@me.com; federica.gabella@gmail.com; guido.tiberio@unibs.it

© The Author(s) 2025
G. A. M. Tiberio (ed.), *Primary Adrenal Malignancies*, Updates in Surgery,
https://doi.org/10.1007/978-3-031-62301-1_12

Surgical debulking aims to control catecholamine-related symptoms and prevent potentially fatal hemodynamic events. Ellis et al. showed that patients with disease located in the abdomen had longer-lasting relief from hypertensive symptoms after debulking. Unfortunately, most patients experienced biochemical recurrence within a year, and 5 years after surgery only 30% of patients had not relapsed. Debulking procedures with evidence of residual disease at the end of surgery resulted in poor and short-lasting biochemical control; upon these bases the authors suggested that a debulking operation should be undertaken only if complete removal of the tumor is anticipated on the basis of preoperative imaging [5].

12.3 Surgical Strategies and Techniques

The surgical strategy for malignant PHEO does not differ from benign disease and must respect the "no touch" or the "as minimal manipulation as possible" principles to avoid hemodynamic instability resulting from catecholamine release. This strategy also prevents tumor effraction and seeding [6, 7].

The surgical approach depends on the surgeon's choice and experience; it can be influenced by patient or tumor characteristics: mass size and location, body mass index (BMI), previous history of abdominal surgery, good control of hypertensive symptoms before surgery.

Minimally invasive surgery can be performed with a transabdominal or a retroperitoneal approach. Most general surgeons prefer the transabdominal route due to the more familiar anatomy and easier conversion to open surgery; the advantage of the retroperitoneal approach is the direct access to the adrenal gland without mobilization of other abdominal organs.

A comparison between transabdominal and retroperitoneal approach showed that in cases treated with the retroperitoneal approach the operative time is shorter, blood loss is lower, and postoperative hospital stay is reduced.

Contraindications to the retroperitoneal approach are tumors larger than 7–8 cm because of the poor working space that can be created in the retroperitoneum, and patients with high BMI because of the large amount of retroperitoneal fat [7].

Laparoscopic adrenalectomy is less invasive as it reduces postoperative pain and allows faster recovery with earlier discharge; it is the first-choice approach when malignancy is unlikely: small tumors with no evidence of local invasiveness or distant metastasis [4].

In malignant PHEOs or in those with a high suspicion of malignancy, the minimally invasive approach is controversial because of the risk of lesion rupture and neoplastic contamination of the peritoneal cavity. In these cases, a complete resection respecting the integrity of the adrenal gland has a critical value. In fact, tumor effraction may result in dissemination of neoplastic cells with pheochromocytomatosis, most often observed at the surgical site or in the surgical wounds including the trocar entry points [8, 9]. A further limitation for minimally invasive surgery is represented by the larger dimensions and higher frequency of infiltration of the tumor capsule and surrounding tissues of malignant PHEOs [6]. Despite this, laparoscopic resection of malignant PHEO proved safe and effective in high-volume centers. In fact, for this disease routine regional adrenalectomy and lymphadenectomy are not required to achieve surgical radicality.

The "vein first" technique was long considered the first step of this surgery. However, this may be difficult to achieve and does not seem to prevent the release of catecholamines. In fact, the venous drainage of the adrenal gland is not carried out only by the main adrenal vein: other "minor" adrenal veins carry blood to the renal capsule, the inferior phrenic veins and even into some tributary veins of the portal system. Some authors have investigated the safety of late adrenal vein ligation and they found that this technique is not associated with increased morbidity or increased frequency of hemodynamic instability during surgery, provided that the procedure is conducted limiting improper manipulation of the gland, which is the main cause of dangerous fluctuations in blood pressure during the procedure [10–14].

The presence of many newly formed vessels reduces the effectiveness of adrenal vein ligation in reducing episodes of hemodynamic instability, the frequency of which depends mostly on the intraoperative manipulation of the mass [2].

Ochi et al. demonstrated that the pre-emptive ligation of the adrenal arteries reduces the amount of catecholamines released from the tumor in the later stages of surgery [12]. Supporters of the retroperitoneal approach also support and disseminate the message of the "artery first" procedure [15].

Otsuka's group reported the case of a patient with a large malignant PHEO with liver invasion. Preoperative computed tomography showed that the tumor received arteries from the celiac tripod and right renal artery. The surgical strategy involved an initial laparoscopic retroperitoneal time during which all the tributary vessels of the tumor found in this space were transected and a second open time that allowed the excision of the mass en bloc with the posterior segments of the liver [16].

12.4 Case Report

In the Supplementary material of this chapter a video is provided concerning the case of a locally advanced adrenal PHEO with multiple nodal metastases. In this case a robot approach was chosen, using indocyanine green dye to detect retroperitoneal lymphatic routes.

References

1. Fassnacht M, Assie G, Baudin E, et al. Adrenocortical carcinomas and malignant phaeochromocytomas: ESMO-EURACAN Clinical Practice Guidelines for diagnosis, treatment and follow-up. Ann Oncol. 2020;31(11):1476–90. Erratum in: Ann Oncol. 2023;34(7):631.
2. AIOM—Associazione Italiana di Oncologia Medica. Linee guida: neoplasie neuroendocrine, 2019 [Neuroendocrine neoplasms guidelines, 2019]. https://www.aiom.it/wp-content/uploads/2019/10/2019_LG_AIOM_Neuroendocrini.pdf.
3. Lenders JW, Duh QY, Eisenhofer G, et al. Pheochromocytoma and paraganglioma: an Endocrine Society clinical practice guideline. J Clin Endocrinol Metab. 2014;99(6):1915–42. Erratum in: J Clin Endocrinol Metab 2023 Apr 13;108(5):e200.

4. Zarnegar R, Kebebew E, Duh QY, Clark OH. Malignant pheochromocytoma. Surg Oncol Clin N Am. 2006;15(3):555–71.
5. Ellis RJ, Patel D, Prodanov T, et al. Response after surgical resection of metastatic pheochromocytoma and paraganglioma: can postoperative biochemical remission be predicted? J Am Coll Surg. 2013;217(3):489–96.
6. Mirica RM, Paun S. Surgical approach in pheochromocytoma. In: Cianci P, Restini E, Agrawal A, editors. Pheochromocytoma, paraganglioma and neuroblastoma. IntechOpen; 2021. https://doi.org/10.5772/intechopen.92492.
7. Patel D. Surgical approach to patients with pheochromocytoma. Gland Surg. 2020;9(1):32–42.
8. Goffredo P, Adam MA, Thomas SM, et al. Patterns of use and short-term outcomes of minimally invasive surgery for malignant pheochromocytoma: a population-level study. World J Surg. 2015;39(8):1966–73.
9. Rafat C, Zinzindohoue F, Hernigou A, et al. Peritoneal implantation of pheochromocytoma following tumor capsule rupture during surgery. J Clin Endocrinol Metab. 2014;99(12):E2681–5.
10. Zografos GN, Farfaras AK, Kassi E, et al. Laparoscopic resection of pheochromocytomas with delayed vein ligation. Surg Laparosc Endosc Percutan Tech. 2011;21(2):116–9.
11. Vassiliou MC, Laycock WS. Laparoscopic adrenalectomy for pheochromocytoma: take the vein last? Surg Endosc. 2009;23(5):965–8.
12. Ochi A, Fan B, Kimura N, et al. Two-step technique of early adrenal artery ligation in open adrenalectomy of giant right adrenal pheochromocytomas: three case reports. IJU Case Rep. 2018;2(1):15–8.
13. Wu G, Zhang B, Yu C, et al. Effect of early adrenal vein ligation on blood pressure and catecholamine fluctuation during laparoscopic adrenalectomy for pheochromocytoma. Urology. 2013;82(3):606–11.
14. Zhang X, Lang B, Ouyang JZ, et al. Retroperitoneoscopic adrenalectomy without previous control of adrenal vein is feasible and safe for pheochromocytoma. Urology. 2007;69(5):849–53.
15. Walz MK, Alesina PF, Wenger FA, et al. Laparoscopic and retroperitoneoscopic treatment of pheochromocytomas and retroperitoneal paragangliomas: results of 161 tumors in 126 patients. World J Surg. 2006;30(5):899–908.
16. Otsuka I, Kida K, Terada N, et al. Malignant pheochromocytoma with liver invasion treated successfully by combined retroperitoneal laparoscopic control of arterial in-flow followed by open hepatectomy: a case report. Int J Surg Case Rep. 2021;81:105763.

Open or Laparoscopic Surgery in the Management of Adrenocortical Carcinoma?

13

Giovanni Emiliani, Silvia Ministrini, Sarah Molfino, and Guido A. M. Tiberio

13.1 Introduction

The best surgical approach for adrenocortical malignancies is still a matter of debate.

In the last 30 years, laparoscopic adrenalectomy (LA) has become the treatment of choice for benign adrenal disorders, both functioning and nonfunctioning [1]. The benefits of the minimally invasive approach are well documented in the literature and consist of improved postoperative recovery, shorter length of hospital stay, lower rate of perioperative complications and reduced cost [2, 3].

For adrenocortical carcinoma (ACC), however, the role of laparoscopy is debated, and the literature is conflicting: some observational studies raise concern regarding the oncologic outcome after a laparoscopic approach while others suggest its safety. Also, in this particular area, the literature is affected by the low incidence of the disease: no randomized trials have been published and the available studies are based on retrospective series recruited over a large time span. A curative resection should provide negative margins, integrity of the tumor capsule, en bloc removal of the tumor with the periadrenal fat and adjacent infiltrated organs, and locoregional lymphadenectomy [4]. A curative (R0) resection must be pursued whenever possible, because it is the most important determinant of long-term survival [5]. For these reasons and because of the complexity of the procedure, open adrenalectomy (OA) is the preferred approach for ACC. However, the spread of laparoscopic techniques and the evolution of the available technology, which now includes robotic assistance, allow for more extensive operations with minimally invasive techniques, keeping the debate alive.

G. Emiliani (✉) · S. Ministrini · S. Molfino · G. A. M. Tiberio
General Surgery, Department of Clinical and Experimental Sciences, University of Brescia at ASST Spedali Civili di Brescia, Brescia, Italy
e-mail: giovanni.emiliani3@gmail.com; ministrini.silvia@me.com; sarahmolfino@gmail.com; guido.tiberio@unibs.it

In this chapter we analyze the available literature to identify the critical elements of the discussion. Location, tumor size and surgeon's experience all contribute to the choice of approach.

13.2 Studies in Favor of Minimally Invasive Adrenalectomy

In a retrospective single-center study, Donatini et al. showed the association of LA with a shorter length of hospitalization without compromising long-term oncological outcomes for stage I–II ACC ≤10 cm [6]. In another retrospective single-center study, Porpiglia et al. found equal oncological outcome between patients subjected to OA and LA for stage I–II ACC when oncologic principles are respected [7]. The same conclusion is reported in a multi-institutional Italian survey by Lombardi et al. [8]. Similarly, Brix et al. described the same postoperative outcomes in OA and LA in terms of survival, capsule rupture and carcinomatosis for ACC stage I–II and even stage III, although the latter was more frequently approached with the open technique [9].

More recently, Maurice et al. used the National Cancer Database to compare outcomes for patients with stage I–IV ACC undergoing OA versus minimally invasive adrenalectomy. Although positive surgical margins were more common in the minimally invasive group, no statistically significant differences in 3-year overall survival were found between the two groups. The authors concluded that minimally invasive adrenalectomy provides acceptable long-term outcomes with faster postoperative recovery for patients with stage I–II ACC [10]. Similarly, Lee et al. retrospectively examined 201 patients from multiple centers. This study found no difference in 30-day mortality rates between the LA group and the OA group. Intraoperative tumor rupture did not occur more frequently in the minimally invasive versus open group, and R0 status was achieved in a comparable number of patients. Parameters such as T stage and not the surgical approach were found to be predictive of survival [11].

The above-mentioned papers in favor of minimally invasive adrenalectomy are summarized in Table 13.1.

Even if these papers clearly state the safety of the LA for early ACC, a deep and critical analysis highlights methodological issues and some biases. In the papers by Donatini et al. and Brix et al., there is a major, statistically significant incidence of smaller tumors in the laparoscopic arms. Similarly, the pre-emptive exclusion of R+ cases from the analysis of data in the papers of Donatini and Lombardi may alter the interpretation. Furthermore, in the studies by Maurice and Lee, statistically significant differences were present in baseline demographics and tumor characteristics. All these anomalies may have biased the results in favor of laparoscopy.

Mpaili et al. reviewed 1171 patients staged ENSAT I–III from 13 studies and concluded that the main point of interest in this discussion is the adequacy of tumor resection rather than the surgical approach itself [12].

Table 13.1 Studies in favor of minimally invasive adrenalectomy

Study	n (LA/OA) years observed	Tumor size (LA/OA)	ENSAT stage (LA/OA)	Tumor rupture (LA/OA)	Positive margin (LA/OA)	Conversion rate	OS (LA/OA)	DFS (LA/OA)	Recurrence rate (LA/OA)	Recurrence pattern	Comments
Donatini (2014) [6] Retrospective single-center study	34 (13/21) years: 1985–2011	55/68 mm p = n.s.	I–II	0/4.7%	0/0%	0%	80 months: 85/81% p = 0.634	Median months: 46/47 p = n.s.	31/24% p = n.s.	NR	Smaller tumors in LA arm
Porpiglia (2010) [7] Retrospective single-center study	43 (18/25) years: 2002–2008	90/105 mm p = n.s.	I: 20/12% II: 80/88% p = n.s.	0/0%	Excluded	0%	Median not reached	Median months: 23/18 p = n.s.	50/64%	Not different	Pre-emptive exclusion of R+ cases
Lombardi (2012) [8] Retrospective multicentric study	156 (30/126) years: 2003–2010	77.3/90.4 mm p = n.s.	I: 33/19% II: 66.7/81% p = n.s.	0/0%	Excluded	0%	Median months: 108/60 p = n.s.	Median months: 72/48 p = n.s.	26.6/38% p = n.s.	Not different	R+ excluded. High rate of incidentalomas in LA arm
Brix (2010) [9] Retrospective multicentric study	152 (35/117) years: 1996–2009	62/80 mm p = 0.001	I: 34/8% II: 55/60% III: 11/32% p = 0.001	9/15% p = n.s.	6/12% p = n.s.	2.8%	NR	NR.	77/69% p = n.s.	Not different	Stage I–II more frequent in LA approach Smaller tumors in LA arm

(continued)

Table 13.1 (continued)

Study	n (LA/OA) years observed	Tumor size (LA/OA)	ENSAT stage (LA/OA)	Tumor rupture (LA/OA)	Positive margin (LA/OA)	Conversion rate	OS (LA/OA)	DFS (LA/OA)	Recurrence rate (LA/OA)	Recurrence pattern	Comments
Maurice (2017) [10] Retrospective study. National Cancer Data Base	481 (161/320) years: 2010–2013	75/117 mm p = 0.01	T1: 54.7/40.9% T2: 15.5/14.7% T3: 24.2/28.1% T4: 5.6/16.3% p = 0.01	NR	20/17% p = n.s.	14.9%	3 years: 62.1/58% p = 0.42	NR	NR	NR	Positive surgical margins more common in LA group, but no statistically significant differences in 3-year OS
Lee (2017) [11] Retrospective multicentric study	201 (47/154) years: 1994–2014	55/109 mm p < 0.001	T1–T2: 75/44.3% T3–T4: 25/55.7% p < 0.001	12.2/9.4 p = n.s.	27.5/28% p = n.s.	19%	5 years: 67.7/48.6% p = n.s.	5 years: 9.1/3.8% p = n.s.	48.9/64.1% p = 0.074	NR	Mini-invasive approach is suggested for ACC <10 cm

DFS disease-free survival, *LA* laparoscopic adrenalectomy, *NR* not reported, *n.s.* not significant, *OA* open adrenalectomy, *OS* overall survival, *R+* not radical resection

13.3   Studies in Favor of Open Adrenalectomy

Some supporters of OA underline the risks of laparoscopy in the management of suspected ACC related to the higher rates of peritoneal carcinosis found in their series [13]. In particular, in the study by Leboulleux et al., no other risk factor for carcinosis (e.g., dimension, stage, functional status, completeness of surgery) was identified except the type of surgical approach. Although these conclusions are likely to be limited by the—at that time—early diffusion of the laparoscopic approach, we must note that the few cases subjected to LA were mostly detected at stage I [14]. In 2018, Wu et al. published a review comprising data on 44 patients who had undergone OA or LA for stage I–II ACC with tumor size less than 10 cm. Local recurrence and peritoneal carcinomatosis trended in favor of OA but the data did not reach statistical significance. However, mean time to local recurrence and peritoneal carcinomatosis was significantly shorter in the LA group compared to the OA group [15].

Other surgeons focused on the higher rate of positive margins with the minimally invasive approach, with consequent worse prognosis and reduction of overall and disease-free survival [16]. In the largest study comparing OA versus LA for suspected ACC, published by Huynh et al. in 2016, 423 patients who had undergone OA or LA for stage I–III ACC were identified from the US National Cancer Center Database. Despite patients in the OA group having larger, more advanced tumors compared to the LA group, LA was identified as an independent risk factor for death on multivariate analysis. Furthermore, margin positivity was higher for T3 tumors treated with LA [17]. Nakanishi et al. recently reviewed 1617 cases from 11 different studies; they demonstrated a lower rate of positive resection margins in favor of OA. The open approach also had better overall and recurrence-free survival rates than laparoscopic surgery at 3 years. Unfortunately, some studies included in this review were of poor quality due to an insufficient follow-up period, so the results appear inconclusive [18].

Some authors examined how the oncological outcome can be influenced by the conversion from a minimally invasive to an open approach and by the tumor size. In a series of 588 patients by Calcatera et al., no difference in median survival was observed between LA and OA, but median survival for the minimally invasive surgery group was twice that for the converted group. Multivariate analysis then showed that size greater than 5 cm was the only predictor of conversion from LA to OA and that size greater than 5 cm, as well as positive margins, were independent predictors of worse overall survival in patients treated with laparoscopic/robotic adrenalectomy. These results appear to suggest that LA may be useful only if it is possible to achieve full resection of the ACC [19].

Finally, the role of the volume-outcome relationship was proposed for ACC by Cooper et al. In this study patients were stratified not only by type of procedure (OA versus LA) but also by location of surgery. Patients referred to the authors' tertiary hospital after OA or LA resection at other hospitals were compared with patients treated with OA resection primarily at the authors' hospital. A higher rate of R0 margins and a lower rate of peritoneal carcinomatosis were recorded when the OA

Table 13.2 Studies in favor of open adrenalectomy

Study	n (LA/OA) years observed	Tumor size (LA/OA)	ENSAT stage (LA/OA)	Tumor rupture (LA/OA)	Positive margin (LA/OA)	Conversion rate	OS (LA/OA)	DFS (LA/OA)	Recurrence rate (LA/OA)	Recurrence pattern (LA/OA)	Comments
Leboulleux (2010) [114] Retrospective cohort study	64 (6/58) years: 2003–2009	70/140 mm p = 0.006	I: 83/3.5% II: 0/46.5% III: 0/12% IV: 17/34.5%	NR	17/36%	NR	NR	NR	NR	PC: 66.7/26.8% p = 0.016	Procedures performed in different centers LA risk factor for PC
Wu (2018) [15] Retrospective single-center study	44 (21/23) years: 2009–2017	58/68.7 mm p = 0.075	I: 28.6/13% II: 71.4/87% p = n.s.	NR	NR	4.7%	5 years: 47/43% p = 0.635	5 years: 39/36% p = 0.802	52/52% p = 0.989	LR and PC 42/22% p = 0.035	Higher risk of LR and PC after LA
Miller (2012) [16] Retrospective single-center study	156 (46/110) years: 2005–2011	74/120 mm	I: 8.7/0% II: 63/50% III: 28.3/50% p = NR	NR	30/16% p = 0.04	NR	Median months: Stage II: 51/103 p = 0.002	Median months: Stage II: 17.6/52.9 p = 0.120	NR	NR	LA in stage II: Higher rate of margin +, tumor spillage, and recurrence; shorter OS

Study	Patients (years)	Tumor size	Stage		LR						Comments
Huynh (2016) [17] Retrospective study. National Cancer Data Base	423 (137/286) years: 2010–2014	80/127 mm p = 0.001	I: 14.6/3.9% II: 54.7/53.2% III: 30.7/43.0% p = n.s.	NR	Overall: 18.3/15.0% p = 0.58 T3: 54.6/21.7% p < 0.0009	NR	NR	NR	NR	NR	LA decreases OS in stage II Nodal assessment: LA 2.9%; OA 30.8%
Calcatera (2018) [19] Retrospective study. National Cancer Data Base	588 (200/388) years: 2010–2014	89/124 mm p = 0.001	T1: 9.5/2.3% T2: 26/1.4% T3: 18/20.1% T4: 5/12.6% unknown: 40/30.9% p = 0.001	NR	18/14.9 p = 0.56	23.6%	Median months: 53/55 p = 0.93	NR	NR	NR	Size greater than 5 cm was the only predictor of conversion
Cooper (2013) [20] Retrospective multicentric study	302 LA 46 OAT 46 OAE 210 years: 1993–2012	LA 80 mm OAT 123 mm OAE 120 mm p < 0.0001	LA/OAT/OAE I: 19.6/4.3/1.4% II: 43.5/52.2/49.5% III: 34.8/21.7/29% IV: 0/21.7/15.7% p = n.s.	NR	LA 21.1% OAT 8.7% OAE 14.3% p = 0.01	8.7%	Median months: LA 53.5 OAT 109.8 OAE 46 p = 0.07	Median months: LA 10.9 OAT 19.6 OAE 9.5 p = 0.005	LA 76.1% OAT 58.7% OAE 87.3% p = 0.001	PC: LA 54.3% OAT 19.6% OAE 27.6% p = 0.006	R2 excluded PC more frequent in LA group

DFS disease-free survival, *LA* laparoscopic adrenalectomy, *LR* local recurrence, *NR* not reported, *n.s.* not significant, *OA* open adrenalectomy, *OAE* open adrenalectomy in external hospitals, *OAT* open adrenalectomy in a tertiary hospital, *OS* overall survival, *PC* peritoneal carcinosis, *R2* macroscopic residual tumor

resection was primarily performed at the tertiary hospital rather than at the referring outside hospitals. It may therefore be supposed that also the type of hospital may influence patient outcomes [20].

Table 13.2 reports the main papers in favor of open adrenalectomy.

13.4   Discussion and Guideline Recommendations

It must be underlined that a major methodological bias has affected the debate and altered the evidence. Enrollment in all these retrospective studies was conducted ex post and based on the pathological diagnosis of ACC and not on the clinical diagnosis, as it should have been. The clinical dilemma surrounding the "undetermined adrenal mass" and the difficulties encountered in the preoperative diagnosis of early-stage ACC heavily impact the conclusions of retrospective studies. In general, only a minority of early-stage ACCs are correctly recognized and staged before surgery; more often the diagnosis is formulated postoperatively by the pathologist. In the co-operative Italian paper, 53% of ACCs were approached with a preoperative diagnosis of incidentaloma (83% in the laparoscopic arm) [8]. The penetration of this phenomenon varies greatly in the different papers and its real impact cannot be recognized owing to the retrospective nature of the studies. Indeed, this is a major bias in the interpretation of data: oncologic outcomes of different surgical techniques are compared across inhomogeneous technical settings with a significant percentage of surgical procedures performed for lesions which are not diagnosed as malignant and, it is reasonable to assume, are not approached with those cautions and technical requirements normally adopted when approaching a malignant lesion. Inevitably, considering the "grey scale" characteristic of the undetermined adrenal mass, those masses that appear at higher risk of malignancy (larger tumors, enlarged lymph nodes, inconclusive imaging) seem to be more often approached with open surgery while those presenting more favorable characteristics are more likely to be treated with minimally invasive techniques.

The above considerations justify the prudential use of the laparoscopic approach suggested by all the guidelines [4, 21–23]. The open approach as the surgical standard of care for confirmed or highly suspected ACC is recommended by: the European Society of Endocrine Surgeons (ESES), European Network for the Study of Adrenal Tumors (ENSAT), European Society for Medical Oncology (ESMO), European Reference Network on Rare Adult Cancers (EURACAN), American Association of Clinical Endocrinologists (AACE), American Association of Endocrine Surgeons (AAES), and Society of Gastrointestinal and Endoscopic Surgeons (SAGES). Furthermore, they suggest the use of the laparoscopic approach only in the absence of local invasion, suspected metastatic lymph nodes and when the principles of oncological surgery can be respected. This surgery should be reserved for expert surgeons and centralized to high-volume institutions. In any case, immediate conversion to open surgery whenever there is a risk of incomplete resection if the operation is conducted laparoscopically is strongly recommended.

References

1. Porpiglia F, Garrone C, Giraudo G, et al. Transperitoneal laparoscopic adrenalectomy: experience in 72 procedures. J Endourol. 2001;15(3):275–9.
2. Yan Y, Cheng J, Chen K, et al. Better clinical benefits and potential cost saving of an enhanced recovery pathways for laparoscopic adrenalectomy. Gland Surg. 2022;11(1):23–34.
3. Hazzan D, Shiloni E, Golijanin D, et al. Laparoscopic vs open adrenalectomy for benign adrenal neoplasm. Surg Endosc. 2001;15(11):1356–8.
4. Fassnacht M, Assie G, Baudin E, et al. Adrenocortical carcinomas and malignant phaeochromocytomas: ESMO-EURACAN Clinical Practice Guidelines for diagnosis, treatment and follow-up. Ann Oncol. 2020;31(11):1476–90. Erratum in: Ann Oncol. 2023;34(7):631.
5. Margonis GA, Kim Y, Prescott JD, et al. Adrenocortical carcinoma: impact of surgical margin status on long-term outcomes. Ann Surg Oncol. 2016;23(1):134–41.
6. Donatini G, Caiazzo R, Do Cao C, et al. Long-term survival after adrenalectomy for stage I/II adrenocortical carcinoma (ACC): a retrospective comparative cohort study of laparoscopic versus open approach. Ann Surg Oncol. 2014;21(1):284–91.
7. Porpiglia F, Fiori C, Daffara F, et al. Retrospective evaluation of the outcome of open versus laparoscopic adrenalectomy for stage I and II adrenocortical cancer. Eur Urol. 2010;57(5):873–8.
8. Lombardi CP, Raffaelli M, De Crea C, et al. Open versus endoscopic adrenalectomy in the treatment of localized (stage I/II) adrenocortical carcinoma: results of a multiinstitutional Italian survey. Surgery. 2012;152(6):1158–64.
9. Brix D, Allolio B, Fenske W, et al. Laparoscopic versus open adrenalectomy for adrenocortical carcinoma: surgical and oncologic outcome in 152 patients. Eur Urol. 2010;58(4):609–15.
10. Maurice MJ, Bream MJ, Kim SP, Abouassaly R. Surgical quality of minimally invasive adrenalectomy for adrenocortical carcinoma: a contemporary analysis using the National Cancer Database. BJU Int. 2017;119(3):436–43.
11. Lee CW, Salem AI, Schneider DF, et al. Minimally invasive resection of adrenocortical carcinoma: a multi-institutional study of 201 patients. J Gastrointest Surg. 2017;21(2):352–62.
12. Mpaili E, Moris D, Tsilimigras DI, et al. Laparoscopic versus open adrenalectomy for localized/locally advanced primary adrenocortical carcinoma (ENSAT I–III) in adults: is margin-free resection the key surgical factor that dictates outcome? A review of the literature. J Laparoendosc Adv Surg Tech A. 2018;28(4):408–14.
13. Gonzalez RJ, Shapiro S, Sarlis N, et al. Laparoscopic resection of adrenal cortical carcinoma: a cautionary note. Surgery. 2005;138(6):1078–86.
14. Leboulleux S, Deandreis D, Al Ghuzlan A, et al. Adrenocortical carcinoma: is the surgical approach a risk factor of peritoneal carcinomatosis? Eur J Endocrinol. 2010;162(6):1147–53.
15. Wu K, Liu Z, Liang J, et al. Laparoscopic versus open adrenalectomy for localized (stage 1/2) adrenocortical carcinoma: experience at a single, high-volume center. Surgery. 2018;164(6):1325–9.
16. Miller BS, Gauger PG, Hammer GD, Doherty GM. Resection of adrenocortical carcinoma is less complete and local recurrence occurs sooner and more often after laparoscopic adrenalectomy than after open adrenalectomy. Surgery. 2012;152(6):1150–7.
17. Huynh KT, Lee DY, Lau BJ, et al. Impact of laparoscopic adrenalectomy on overall survival in patients with nonmetastatic adrenocortical carcinoma. J Am Coll Surg. 2016;223(3):485–92.
18. Nakanishi H, Miangul S, Wang R, et al. ASO visual abstract: open versus laparoscopic surgery in the management of adrenocortical carcinoma: a systematic review and meta-analysis. Ann Surg Oncol. 2023;30(2):1006–7.
19. Calcatera NA, Hsiung-Wang C, Suss NR, et al. Minimally invasive adrenalectomy for adrenocortical carcinoma: five-year trends and predictors of conversion. World J Surg. 2018;42(2):473–81.
20. Cooper AB, Habra MA, Grubbs EG, et al. Does laparoscopic adrenalectomy jeopardize oncologic outcomes for patients with adrenocortical carcinoma? Surg Endosc. 2013;27(11):4026–32.

21. Gaujoux S, Mihai R. European Society of Endocrine Surgeons (ESES) and European Network for the Study of Adrenal Tumours (ENSAT) recommendations for the surgical management of adrenocortical carcinoma. Br J Surg. 2017;104(4):358–76.
22. Zeiger MA, Thompson GB, Duh QY, et al. The American Association of Clinical Endocrinologists and American Association of Endocrine Surgeons medical guidelines for the management of adrenal incidentalomas. Endocr Pract. 2009;15(Suppl 1):1–20.
23. Stefanidis D, Goldfarb M, Kercher KW, et al. SAGES guidelines for minimally invasive treatment of adrenal pathology. Surg Endosc. 2013;27(11):3960–80.

Pathology of Adrenocortical Carcinoma and Malignant Pheochromocytoma

14

Giulia Vocino Trucco and Marco Volante

14.1 Introduction

Adrenocortical carcinoma (ACC) and pheochromocytoma (PHEO) are the two most frequent types of malignant primary tumors of the adrenal gland. ACC is classified within the spectrum of adrenocortical tumors that also include adrenocortical adenoma and adrenocortical nodular disease, the latter being a group of lesions that are now considered neoplasms, but were previously termed primary forms of adrenocortical hyperplasia. By contrast, PHEO originates from chromaffin cells in the adrenal medulla, belongs to the family of sympathetic paragangliomas, and—in line with the general concept of classification for neuroendocrine neoplasms—is malignant by definition, although most cases have an indolent clinical course.

Pathologically, the diagnosis of both ACC and PHEO is not straightforward, and the evaluation of several pathological parameters—mostly coded into scoring systems—is needed to define the risk of clinical malignancy. In the present chapter, we will summarize the most relevant pathological findings of ACC and PHEO, focusing on diagnostic algorithms and clinically useful tissue biomarkers; the pathology flow-chart is reproduced in Fig. 14.1.

G. Vocino Trucco
Pathology Unit, ASL CN 1—SS. Annunziata Hospital, Savigliano (Cuneo), Italy
e-mail: giulia.vocinotrucco@aslcn1.it

M. Volante (✉)
Department of Oncology, University of Turin, San Luigi Gonzaga University Hospital, Orbassano (Turin), Italy
e-mail: marco.volante@unito.it

© The Author(s) 2025
G. A. M. Tiberio (ed.), *Primary Adrenal Malignancies*, Updates in Surgery,
https://doi.org/10.1007/978-3-031-62301-1_14

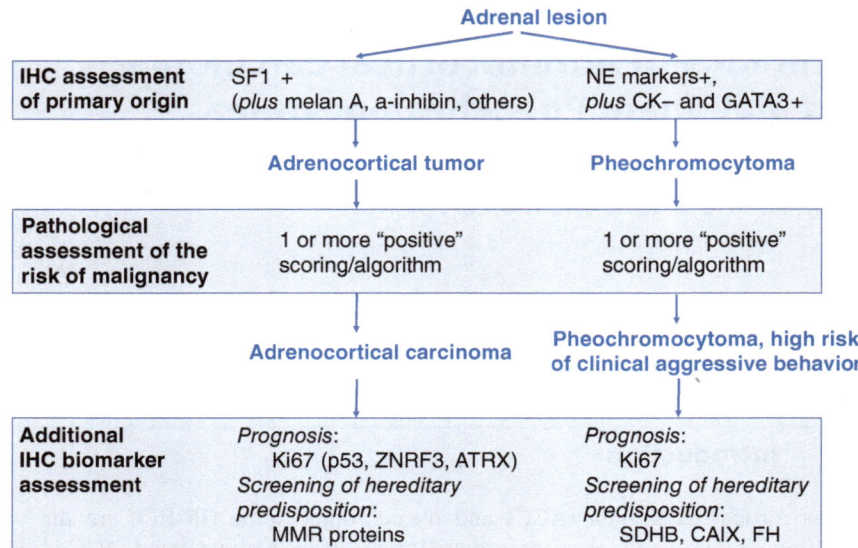

Fig. 14.1 Pathology flow-chart for adrenocortical carcinoma and pheochromocytoma. *IHC* immunohistochemical

14.2 Adrenocortical Carcinoma

14.2.1 Gross Pathology

The macroscopic appearance of ACC is extremely heterogeneous and may present either an encapsulated lesion or a mass infiltrating surrounding structures. ACC is usually large, with a mean size of about 11 cm and an average weight of about 400 g [1]. It usually loses the yellowish homogeneous cut surface typical of adrenocortical adenoma, and most often combines a more whitish-to-grayish appearance, with a more or less prominent stromal component and hemorrhagic and/or necrotic areas. The overall appearance is usually suggestive of malignancy and, more than with adrenocortical adenoma, it requires a differential diagnosis with other malignant tumors in the adrenal gland, such as other malignant primaries (aggressive forms of PHEO or sarcomas) or metastatic tumors. However, some cases are smaller, and their size and weight are not indicative of malignancy. Small lesions show a more homogeneous and less suspicious appearance, with clear demarcation and no apparent invasion of capsular and extracapsular structures; if adequate sampling is not performed, these cases may risk being underdiagnosed as benign.

More than being informative *per se* in supporting ACC diagnosis, an accurate macroscopic description and evaluation is key to guiding a precise and exhaustive sampling procedure and correctly defining the status of the resection margins.

14.2.2 Cytological and Histological Findings

Cytologically, ACC usually contains a predominance of lipid-depleted cells with a dense eosinophilic cytoplasm. The cytoplasmic features of ACC cells do not correlate with any specific endocrine syndrome nor are they associated with specific functional properties. Nuclear pleomorphism and the presence of nucleoli are almost always a prominent feature in ACC, some cases having a very high degree of atypia. However, nuclear atypia may also occur in benign adrenocortical lesions and is therefore a fairly non-specific feature.

The tumor architecture in ACC is frequently heterogeneous, with different growth patterns frequently coexisting within the same lesion. Irrespective of the tumor architecture, a consistent finding in ACC is the loss of the well-organized alveolar pattern seen in non-tumorous adrenocortical tissue and in adrenocortical adenoma, and this observation is the key item of one of the algorithms for ACC diagnosis, the reticulin algorithm [2] (see below). Characteristic architectural patterns in ACC include a broad trabecular growth, with anastomosing columns and cords of cells, a nesting or alveolar arrangement or a more diffuse or solid growth with a pattern-less histological architecture. Such heterogeneity is a relevant issue in the histological differential diagnosis of ACC. Uncommon architectural patterns include pseudopapillary and storiform, the latter typical of the sarcomatoid variant.

Necrosis, either punctate or extensive, is frequent and should be kept separate from ischemic-type necrosis, which may occur as a consequence of fine-needle aspiration biopsy procedures in benign lesions.

The mitotic c index is usually elevated in ACC, accepting the fact that—as for other endocrine tumors—mean mitotic activity in malignant lesions is generally lower than in other malignancies (i.e., lung, breast, or colon cancer). An elevated mitotic index is probably the most specific feature of malignancy and is incorporated in all scoring systems or diagnostic algorithms, using the same cut-off of >5 mitoses per 10 mm^2 (50 high power fields). However, the distribution of mitotic figures is heterogeneous and they usually cluster in hot-spots; therefore, the evaluation of different fields within the same slides and/or of different slides is advisable. Atypical mitotic figures are suggestive of abnormal chromosome content (aneuploidy) and, when present, represent a hallmark of malignancy, even when a single but unequivocal mitotic figure is identified.

Invasive properties in ACC include capsular, vascular and sinusoidal invasion. Capsular invasion is defined as complete penetration of the tumor capsule, although it can be difficult to evaluate in cases where the tumor capsule is irregular and distorted by fibrous septa. Vascular invasion is another parameter highly specific of malignancy, but it may be missed in a relevant proportion of cases. If recorded using stringent criteria, it was also shown to be an adverse predictor of metastatic disease and clinical outcome [3]. Sinusoidal invasion is equivocally considered either as the presence of tumor cells in thin-walled vascular spaces within the tumor or—more consistent with current guidelines for pathology reporting [4]—as the invasion of lymphatic vessels at the periphery of the tumor.

14.2.3 Scoring Systems

The diagnosis of ACC is based on the combination of architectural and cytological features, of necrosis and mitotic activity and of the above-mentioned invasion-related parameters. At variance with other endocrine neoplasms (i.e., thyroid tumors), no parameter is reliable enough to code malignancy, but all of them have to be considered and assessed individually, and the final diagnosis results from the application of specific scoring systems or algorithms endorsed by the WHO classification.

Strengths of such an approach are the definition of diagnostic rules, a comprehensive pathology report describing all the relevant pathological parameters, and the assessment in some cases of a quantitative evaluation that is also relevant for prediction of clinical outcome. The systems/scorings proposed by the WHO classification of endocrine and neuroendocrine tumors [5] are detailed in Table 14.1. No single system has been shown to be completely sensitive or specific in all settings, nor has any of them proven to have complete observational concordance in individual lesions. Therefore, the WHO continues to suggest the use of multiple approaches to describe the lesion and predict its clinical behavior along with the recording of data for future validation studies. In general terms, it is advisable that pathologists use their judgment to select the appropriate system or multiple systems according to the morphological features of the single lesion they are observing.

The Weiss score is the first scoring system for assessing malignancy described in ACC and is by far the most widely adopted and validated [6]. It consists of nine parameters, and malignancy is defined by the presence of ≥3 positive parameters (range 0 to 9). The Helsinki score is more recent and in principle it has been designed to simplify the Weiss score by limiting the number of variables (thus improving reproducibility), and to integrate Ki67 as an additional tool [7]. Two pathological parameters (mitotic index and necrosis) are considered with different statistical power, together with the exact value of the Ki67 proliferation index (as %). A Helsinki score >8.5 points is associated with metastatic potential with 100% sensitivity and 99.4% specificity. The Helsinki score has been largely evaluated and validated in independent series [8] for both conventional ACC and ACC variants and it has been shown to outperform the Weiss system [9].

The reticulin algorithm proposal stems from evidence that the vascular network is almost invariably regular in adrenocortical adenoma, closely mimicking the normal adrenal cortex, whereas it is disrupted at a variable degree of distribution and quality in ACC. This difference is easily highlighted by silver-based staining procedures, both in terms of qualitative and quantitative changes [1, 2]. To increase specificity for malignancy, in the algorithm the presence of a disrupted reticulin framework should be associated with one or more among the following: increased mitotic index (same cut-off as for the Weiss and Helsinki scores), necrosis, and vascular invasion. The diagnostic performance of the reticulin algorithm has been validated in several studies and endorsed by the WHO classification [5].

Table 14.1 Pathological parameters used for adrenocortical carcinoma diagnosis in the different scoring systems

Parameter	Weiss score	Helsinki score	Reticulin algorithm	Lin-Weiss-Bisceglia system[a]
Nuclear grade 3–4[b]	1 point	–	–	–
Clear cell cytoplasm <25%	1 point	–	–	–
Diffuse architecture >30%	1 point	–	–	–
Mitotic index >5 in 10 mm^2	1 point	3 points	Additional parameter	Major criterion
Atypical mitotic figures	1 point	–	–	Major criterion
Necrosis	1 point	5 points	Additional parameter	Minor criterion
Vascular invasion	1 point	–	Additional parameter	Major criterion
Capsular invasion	1 point	–	–	Minor criterion
Sinusoidal invasion	1 point	–	–	Minor criterion
Disruption of reticulin framework	–	–	Main parameter	–
Ki67 index (as %)	–	Value of Ki67 index	–	–
Size >10 cm	–	–	–	Minor criterion
Weight >200 g	–	–	–	Minor criterion
Cut-off/rule for malignancy	*Score ≥3*	*Score >8.5*	*Presence of the main parameter (altered reticulin pattern) + at least one of the three additional parameters*	*Malignant: presence of at least one of the major criteria Uncertain malignant potential: presence of one or more minor criteria only*

[a] Specifically developed for oncocytic adrenocortical tumors
[b] According to Fuhrman grading of renal cell carcinoma

For predominant oncocytic tumors (see below), the Helsinki score and reticulin algorithm approaches are both indicated, but the Weiss score has a high risk of overestimating malignancy due to the presence of three parameters linked to the finding of oncocytic cells irrespective of their biological nature (i.e., eosinophilic cytoplasm, nucleoli and diffuse growth). Therefore, about 20 years ago, a group of pathologists designed an alternative system based on major (to define malignant cases) and minor (to define cases with uncertain malignant potential) criteria [10].

14.2.4 Histological Subtypes

The heterogeneity of histological features corresponds to the presence of histological variants. Apart from conventional ACC, three main variants are encountered in the WHO classification, namely—in order of decreasing frequency—oncocytic, myxoid and sarcomatoid [11].

Oncocytic ACC represent about 25% of cases, with clinical characteristics, in terms of epidemiology and functional properties, similar to conventional ACC apart from a higher prevalence of functioning tumors secreting sex steroid hormones [12]. Oncocytic ACC histologically is characterized by the presence of a predominant population of oncocytes, whereas cases with a lesser component of oncocytic cells are included in the conventional type. As for oncocytes in other pathological conditions, tumor cells in oncocytic ACC are characterized by abundant eosinophilic and granular cytoplasm, enlarged nuclei with nucleoli and a diffuse growth pattern. The ultrastructural hallmark is the presence of an abnormal accumulation of functionally defective and morphologically altered mitochondria. Myxoid ACC has been only recently recognized as a specific ACC variant. It is rarer than the oncocytic type, but prevalence data in large series are still missing. Histologically, they are characterized by small uniform cells, with mild atypia, embedded in an abundant extracellular matrix made of myxoid material [13]. Sarcomatoid ACC is very rare, with few cases reported in the literature [14]. Most cases have biphasic morphology, with an epithelioid component of conventional type admixed to a variable extent with a sarcomatoid pattern, characterized by spindle cell architecture and highly pleomorphic cells. When monophasic, sarcomatoid ACC can be indistinguishable from true sarcomas of the adrenal gland.

14.2.5 Pediatric Adrenocortical Tumors

Pediatric adrenocortical tumors deserve separate mention due to their peculiar clinical characteristics and pathological features. They more frequently occur before the age of 5 years, with a second peak in adolescence, and females are more frequently affected [15]. Functional tumors are more frequent than in adults, accounting for up to 85% of cases, and virilization is the most frequent functional manifestation. The pathological classification is challenging since the inconsistency of the correlation between tumor behavior and histopathological findings is more pronounced than in adult cases [16]. The Weiss score has been shown to have good specificity for detecting cases with an aggressive clinical course [17], but it generally overestimates malignant potential in pediatric tumors. In 2003, a new proposal was made by the Armed Forces Institute of Pathology (AFIP) which includes the following parameters: tumor weight >400 g, tumor size >10.5 cm, extra-adrenal extension, invasion into the vena cava, venous invasion, capsular invasion, tumor necrosis, mitotic index >15/20 high power fields (HPF) and atypical mitotic figures [18]. The presence of 0 to 2 criteria defines a lesion as benign, a score of >3 is indicative of malignancy, whereas a score of 3 classifies a lesion as indeterminate for malignancy. As for the other scoring systems/algorithms used in adults, the Helsinki score has been demonstrated to possess a high specificity for malignancy [19]. Moreover, the reticulin algorithm has been recently shown to be easy to apply and the most sensitive histopathological approach to identify aggressive behavior in pediatric adrenocortical tumors [20].

14.2.6 Grading

Grading of ACC was called for several years ago [21] but finally adopted by the AJCC TNM staging system eighth edition and more recently by the WHO classification [5]. Despite the wide range of morphological parameters assessed for diagnostic scoring systems, the mitotic index is the only parameter considered for grading. Differently from the diagnostic cut-off, ACC is segregated into low- or high-grade based on a cut-off of 20 mitoses/10 mm^2).

14.2.7 Staging

Current TNM staging of ACC (Table 14.2) is based on integration of the AJCC eighth edition of TNM system and the staging proposal by the European Network for the Study of Adrenal Tumors (ENSAT) [22]. Pathological stage T1 and T2 ACC are tumors limited to the adrenal gland (≤5 cm or >5 cm, respectively), whereas T3 tumors show invasion of the surrounding tissues and T4 invasion of adjacent organs or vena cava or renal vein. N1 and M1 are defined by any type of lymph node or distant involvement, respectively, irrespective of the number of lesions and site.

In pediatric ACC the staging system was developed by the International Pediatric Adrenocortical Tumor Registry (IPACTR) and the Children's Oncology Group (COG) and include resection margins, presence of metastases, and weight [23].

14.2.8 Immunohistochemical Profile

Immunohistochemical profiling is an essential complement for ACC characterization. It is needed to prove the adrenocortical nature of the lesion, to assist the definition of malignancy, to assess the expression of prognostic markers and to screen for the presence of hereditary predisposition.

Table 14.2 European Network for the Study of Adrenal Tumors (ENSAT) staging system for adrenocortical carcinomas

ENSAT stage	Definition
I	T1, N0, M0
II	T2, N0, M0
III	T1–T2, N1, M0
	T3–T4, N0–N1, M0
IV	T1–T4, N0–N1, M1

T1 tumor ≤5 cm, *T2* tumor >5 cm, *T3* tumor infiltration into surrounding tissue, *T4* tumor invasion into adjacent organs or venous tumor thrombus in vena cava or renal vein, *N0* no positive lymph nodes, *N1* presence of positive lymph nodes, *M0* no distant metastases, *M1* presence of distant metastases

For the definition of primary adrenocortical origin, the use of a panel of markers is advisable to balance the sensitivity and specificity of each single marker. ACC is usually negative or only focally positive for cytokeratins, a clue that may help to distinguish ACC from metastatic carcinomas. The best marker for the definition of adrenocortical primary origin is SF1, which is considered the most reliable and specific. It is expressed in steroidogenic cells of the gonads, in the gonadotrophs of the pituitary gland and in normal adrenal cortex and all types of adrenocortical neo-plasms [24]. SF1 specificity for adrenocortical origin is up to 100% and its sensitiv-ity is about 95%, especially if monoclonal antibodies are used [25]. Other markers to be considered in the panel are melan-A, synaptophysin, alpha-inhibin and cal-retinin [26]. However, these markers are also positive in other tumors that may involve the adrenal, and a positive result should be interpreted in relation to the overall pathological picture.

For the definition of malignancy, the use of phospho-histone-H3 immunohisto-chemistry may help to highlight mitoses, especially in carcinomas with low mitotic counts [27]. Other markers have also been proposed to assist in the distinction between adrenocortical adenoma and ACC (i.e., IGF2 and beta-catenin), but their real clinical usefulness is still a matter of debate. Among them, p53 expression is often altered (overexpressed or lost) in the presence of *TP53* mutations. Altered p53 expression has been widely used to support the diagnosis of ACC but not all ACC have *TP53* muta-tions, thus undermining its sensitivity. More interestingly, p53 altered expression has been also demonstrated to be associated with a poorer prognosis [28].

Ki67 *per se* is a diagnostic marker now integrated in the Helsinki score. However, its major role is linked to its strong and independent prognostic impact [29]. Consensus on prognostic cut-offs has not been reached and both three-tier and two-tier categorization approaches are described. Moreover, reproducibility and stan-dardization of Ki67 index evaluation is rather poor [30].

Molecular characterization of ACC has improved in recent years, and transcrip-tome and pan-genomic studies clearly underpinned the role of molecular risk strati-fication in ACC. However, this evidence lacks translation and validation into well-established biomarkers to be assessed in routine clinical work, with special reference to immunohistochemical biomarkers assessable in every pathology labo-ratory. Among them, the loss of ATRX and ZNRF3 expression [31] reflects the presence ACC-specific molecular alterations, and has been proposed to be associ-ated with a more aggressive biological behavior, but the use of such markers in clinical practice still needs validation.

Finally, mismatch-repair (MMR) proteins and PD-L1 immunohistochemistry may also help to determine the eligibility for immunotherapy in select patients, although these markers have no clear indication to date, nor have they been shown to be definitely associated with patient response in large studies [32]. However, the WHO strongly recommend testing for MMR proteins in all ACC patients to screen for the presence of hereditary cases linked to the Lynch syndrome, as indi-cated for other cancer types (i.e., endometrial or colon cancer). Indeed, ACC is part of the Lynch syndrome phenotype and up to 10% of ACC cases are inherited in this context [33].

14.3   Malignant Pheochromocytoma

14.3.1 Gross Pathology

The most common appearance of a sporadic PHEO is a solitary, round or oval mass that distorts the adrenal gland. Tumor size and weight can vary widely, with tumors measuring more than 10 cm and weighting up to several 100 g. On section, the tumor is usually well circumscribed and may even appear encapsulated. PHEOs more commonly have a variegated tan brown and red soft cut surface with areas of frank hemorrhage and central degeneration with necrosis, fibrosis or cystic change. All such features are not indicative alone of a malignant clinical behavior. Multiple nodules on a background of diffusely expanded medulla are suggestive of a hereditary predisposition.

14.3.2 Cytological and Histological Findings

The neoplastic cells are usually large and polygonal with abundant granular basophilic or eosinophilic cytoplasm and indistinct cell borders. The nuclei are round to oval, with prominent nucleoli and nuclear inclusions. Some tumors exhibit significant nuclear pleomorphism with occasional multinucleation.

PHEOs have a characteristic architecture of well-defined nests of tumor cells known as "zellballen", surrounded by a thin fibrovascular stroma and non-neoplastic sustentacular cells. The periphery of the lesion is usually delineated from the surrounding gland, but most tumors do not have a well-defined capsule or pseudocapsule, and it is not uncommon to see intermingling tumor cells and cortical cells. There may be areas of degeneration associated with hemorrhage, fibrosis, and hemosiderin deposition, but coagulative necrosis is unusual. As for other neuroendocrine neoplasms, melanin pigment can be found and, when abundant, may be a worrisome feature in the differential diagnosis with other pigmented lesions such as melanoma. Some pathological features are suggestive of a specific pathogenetic mechanism and, although not completely specific, may suggest a hereditary context. The presence of a myxoid, hyalinized stroma with edema together with enlarged cytoplasm with lipid vacuoles should prompt consideration of von Hippel-Lindau (VHL), whereas tumors with succinate dehydrogenase (SDH) mutations may be composed of large cells with particularly abundant eosinophilic granular cytoplasm. The presence of multifocal disease is also strongly suggestive of genetic predisposition and the identification of adrenal medullary hyperplasia raises the possibility of MEN 2.

14.3.3 Pathological Prediction of a Clinically Aggressive Course

As stated, PHEO is a tumor with potentially malignant biological behavior by definition. Clinical malignancy is certain in the presence of metastases. However, the diagnosis of metastatic disease must be made with great caution, especially in patients with known germline predisposition who are likely to develop multifocal disease in various locations, including visceral organs (such as the liver and lungs). Exceptions are the brain, bone, and lymph nodes, and these are the only sites that can be accepted *a priori* as metastatic locations, if confirmed histologically.

A number of histologic scoring schemes have been developed to predict clinical malignancy in PHEO (Table 14.3), but, unlike those used in ACC, they are not definitely validated or sufficiently accurate to be endorsed by the WHO classification [34]. The oldest score, which applies to PHEO but not to paraganglioma, is the PASS (Pheochromocytoma of the Adrenal gland Scaled Score). This includes 12, differently weighted, histologic parameters [35]. The advantage of the scheme is that it is only based on morphological parameters and does not incorporate clinical data, which are frequently not available to pathologists. The PASS scheme is an excellent rule-out model, with a low PASS score indicating a non-metastatic clinical course, whereas it is poorly specific [36] and affected by high interobserver variability [37]. The GAPP (Grading system for Adrenal Pheochromocytoma and Paraganglioma) score was developed as an alternative and includes a series of

Table 14.3 Pathological parameters used for predicting clinical malignancy in pheochromocytoma in the different scoring systems

Parameter	PASS	GAPP	COPPS
Large nests/diffuse growth	2 points	1 point	–
Central/confluent necrosis	2 points	2 points	5 points
High cellularity	2 points	1 or 2 points[a]	–
Cell monotony	2 points	–	–
Tumor cell spindling	2 points	–	–
Mitotic index >3 in 2 mm^2	2 points	–	–
Atypical mitoses	2 points	–	–
Extension to adipose tissue	2 points	–	–
Vascular invasion	1 point	1 point	1 point
Capsular invasion	1 point	–	–
Profound nuclear pleomorphism	1 point	–	–
Nuclear hyperchromasia	1 point	–	–
Ki67 index	–	1 or 2 points[b]	–
Type of catecholamine secretion	–	1 point[c]	–
Tumor size >7 cm	–	–	1 point
PS100 loss	–	–	2 points
SDHB loss	–	–	1 point
Cut-off/rule for malignancy	*Score ≥4*	*WD: score 0–2* *MD: score 3–6* *PD: score 7–10*	*Score ≥3*

PASS Pheochromocytoma of the Adrenal gland Scaled Score, *GAPP* Grading system for Adrenal Pheochromocytoma and Paraganglioma, *COPPS* Composite Pheochromocytoma/paraganglioma Prognostic Score, *WD* well differentiated, *MD* moderately differentiated, *PD* poorly differentiated

[a] 150–250 cells in 10 mm^2 at 400× = 1 point; >250 cells in 10 mm^2 at 400× = 2 points

[b] 1–3% = 1 point; >3% = 2 points

[c] Adrenergic type (epinephrine ± norepinephrine) = 1 point. Noradrenergic type (norepinephrine ± dopamine) = 0 points. Non-functioning = 0 points

histologic criteria, the Ki67 labeling index, and the patient's biochemical catechol-amine status [38]. According to this system, lesions are stratified into a three-tier scheme of "well-differentiated", "moderately differentiated", and "poorly differentiated", a terminology that is not endorsed by the WHO classification. More recently, the COPPS score (Composite Pheochromocytoma/paraganglioma Prognostic Score) has been proposed that incorporates tumor size, necrosis, vascular invasion and immunohistochemical biomarkers [39] but has not been widely validated.

14.3.4 Staging

Despite the current viewpoint that all pheochromocytomas/paragangliomas (PPGLs) may exhibit metastatic potential, the first PPGL staging system was introduced in the eighth edition of the AJCC Cancer Staging Manual. It is based on tumor local-ization, tumor size, and presence or absence of regional or distant metastases. This staging system is only applied to sympathetic paragangliomas and PHEOs and has been validated in large retrospective series [40] but requires further validation stud-ies to be used in clinical practice.

14.3.5 Immunohistochemical Profile

PHEO has a characteristic immunohistochemical profile [41]. The tumor cells are diffusely positive for common neuroendocrine markers, such as INSM1, syn-aptophysin and chromogranin A. Due to their nonepithelial origin, they are nega-tive for cytokeratins, a characteristic that may assist in the differential diagnosis with other neuroendocrine neoplasms. In this context, another relevant marker is GATA3, whereas functional markers such as tyrosine hydroxylase are also spe-cific but usually not accessible in routine diagnostic laboratories. S100/SOX10 sustentacular cells may be present also in other neuroendocrine neoplasms and are not specific for a chromaffin cell origin. However, a positive S100 and SOX10 immunohistochemistry may be used to rule out a metastatic lesion, since susten-tacular cells are absent in distant metastases. Neuroendocrine hormones secreted by PHEO cells are a variety, but they are usually not used for diagnostic pur-poses, if not to characterize specific cases with unexpected functional properties (i.e., ACTH-secreting cases associated with Cushing's syndrome) [42]. As for other neuroendocrine neoplasms, Ki67 labeling is used to define the proliferative rate, but unlike in the gastroenteropancreatic system it is not indicative of a spe-cific grade, although a high proliferation rate has been shown to be associated with a worse clinical outcome [43]. A major role of immunohistochemistry is to guide assessment of genetic predisposition [44]. The most widely used tool is immunostaining for SDHB. This test is useful because it has been found that mutation in any SDH-related gene (*SDHA, SDHB, SDHC, SDHD* or *SDH-AF2*) results in a loss of SDHB expression in tumor cells. Other immunohistochemical reactions may be surrogate markers of hereditary predisposition. For example,

CAIX expression has been associated with VHL-associated disease, and a loss of FH immunoreactivity may help to identify rare cases of FH-mutated PHEOs, together with intact staining for 2-succinocysteine (2SC).

References

1. Duregon E, Fassina A, Volante M, et al. The reticulin algorithm for adrenocortical tumor diagnosis: a multicentric validation study on 245 unpublished cases. Am J Surg Pathol. 2013;37(9):1433–40.
2. Volante M, Bollito E, Sperone P, et al. Clinicopathological study of a series of 92 adrenocortical carcinomas: from a proposal of simplified diagnostic algorithm to prognostic stratification. Histopathology. 2009;55(5):535–43.
3. Mete O, Gucer H, Kefeli M, Asa SL. Diagnostic and prognostic biomarkers of adrenal cortical carcinoma. Am J Surg Pathol. 2018;42(2):201–13.
4. Giordano TJ, Berney D, de Krijger RR, et al. Data set for reporting of carcinoma of the adrenal cortex: explanations and recommendations of the guidelines from the International Collaboration on Cancer Reporting. Hum Pathol. 2021;110:50–61.
5. Mete O, Erickson LA, Juhlin CC, et al. Overview of the 2022 WHO classification of adrenal cortical tumors. Endocr Pathol. 2022;33(1):155–96.
6. Weiss LM, Medeiros LJ, Vickery AL Jr. Pathologic features of prognostic significance in adrenocortical carcinoma. Am J Surg Pathol. 1989;13(3):202–6.
7. Pennanen M, Heiskanen I, Sane T, et al. Helsinki score—a novel model for prediction of metastases in adrenocortical carcinomas. Hum Pathol. 2015;46(3):404–10.
8. Duregon E, Cappellesso R, Maffeis V, et al. Validation of the prognostic role of the "Helsinki Score" in 225 cases of adrenocortical carcinoma. Hum Pathol. 2017;62:1–7.
9. Minner S, Schreiner J, Saeger W. Adrenal cancer: relevance of different grading systems and subtypes. Clin Transl Oncol. 2021;23(7):1350–7.
10. Bisceglia M, Ludovico O, Di Mattia A, et al. Adrenocortical oncocytic tumors: report of 10 cases and review of the literature. Int J Surg Pathol. 2004;12(3):231–43.
11. de Krijger RR, Papathomas TG. Adrenocortical neoplasia: evolving concepts in tumorigenesis with an emphasis on adrenal cortical carcinoma variants. Virchows Arch. 2012;460(1):9–18.
12. Kanitra JJ, Hardaway JC, Soleimani T, et al. Adrenocortical oncocytic neoplasm: a systematic review. Surgery. 2018;164(6):1351–9.
13. Papotti M, Volante M, Duregon E, et al. Adrenocortical tumors with myxoid features: a distinct morphologic and phenotypical variant exhibiting malignant behavior. Am J Surg Pathol. 2010;34(7):973–83.
14. Papathomas TG, Duregon E, Korpershoek E, et al. Sarcomatoid adrenocortical carcinoma: a comprehensive pathological, immunohistochemical, and targeted next-generation sequencing analysis. Hum Pathol. 2016;58:113–22.
15. Zambaiti E, Duci M, De Corti F, et al. Clinical prognostic factors in pediatric adrenocortical tumors: a meta-analysis. Pediatr Blood Cancer. 2021;68(3):e28836.
16. Dehner LP, Hill DA. Adrenal cortical neoplasms in children: why so many carcinomas and yet so many survivors? Pediatr Dev Pathol. 2009;12(4):284–91.
17. Riedmeier M, Thompson LDR, Molina CAF, et al. Prognostic value of the Weiss and Wieneke (AFIP) scoring systems in pediatric ACC—a mini review. Endocr Relat Cancer. 2023;30(4):e220259.
18. Wieneke JA, Thompson LD, Heffess CS. Adrenal cortical neoplasms in the pediatric population: a clinicopathologic and immunophenotypic analysis of 83 patients. Am J Surg Pathol. 2003;27(7):867–81.
19. Jangir H, Ahuja I, Agarwal S, et al. Pediatric adrenocortical neoplasms: a study comparing three histopathological scoring systems. Endocr Pathol. 2023;34(2):213–23.

20. Lopez-Nunez O, Virgone C, Kletskaya IS, et al. Diagnostic utility of a modified reticulin algorithm in pediatric adrenocortical neoplasms. Am J Surg Pathol. 2024;48(3):309–16.
21. Giordano TJ. The argument for mitotic rate-based grading for the prognostication of adrenocortical carcinoma. Am J Surg Pathol. 2011;35(4):471–3.
22. Fisher SB, Habra MA, Chiang YJ, et al. Comparative performance of the 7th and 8th editions of the American Joint Committee on Cancer staging manual for adrenocortical carcinoma. World J Surg. 2020;44(2):544–51.
23. Lam AK. Adrenocortical carcinoma: updates of clinical and pathological features after renewed World Health Organisation classification and pathology staging. Biomedicines. 2021;9(2):175.
24. Sbiera S, Schmull S, Assie G, et al. High diagnostic and prognostic value of steroidogenic factor-1 expression in adrenal tumors. J Clin Endocrinol Metab. 2010;95(10):E161–71.
25. Duregon E, Volante M, Giorcelli J, et al. Diagnostic and prognostic role of steroidogenic factor 1 in adrenocortical carcinoma: a validation study focusing on clinical and pathologic correlates. Hum Pathol. 2013;44(5):822–8.
26. Mete O, Asa SL, Giordano TJ, et al. Immunohistochemical biomarkers of adrenal cortical neoplasms. Endocr Pathol. 2018;29(2):137–49.
27. Duregon E, Molinaro L, Volante M, et al. Comparative diagnostic and prognostic performances of the hematoxylin-eosin and phospho-histone H3 mitotic count and Ki-67 index in adrenocortical carcinoma. Mod Pathol. 2014;27(9):1246–54.
28. Hescot S, Faron M, Kordahi M, et al. Screening for prognostic biomarkers in metastatic adrenocortical carcinoma by tissue micro arrays analysis identifies P53 as an independent prognostic marker of overall survival. Cancers (Basel). 2022;14(9):2225.
29. Beuschlein F, Weigel J, Saeger W, et al. Major prognostic role of Ki67 in localized adrenocortical carcinoma after complete resection. J Clin Endocrinol Metab. 2015;100(3):841–9.
30. Papathomas TG, Pucci E, Giordano TJ, et al. An international Ki67 reproducibility study in adrenal cortical carcinoma. Am J Surg Pathol. 2016;40(4):569–76.
31. Brondani VB, Lacombe AMF, Mariani BMP, et al. Low protein expression of both ATRX and ZNRF3 as novel negative prognostic markers of adult adrenocortical carcinoma. Int J Mol Sci. 2021;22(3):1238.
32. Remde H, Schmidt-Pennington L, Reuter M, et al. Outcome of immunotherapy in adrenocortical carcinoma: a retrospective cohort study. Eur J Endocrinol. 2023;188(6):485–93.
33. Raymond VM, Everett JN, Furtado LV, et al. Adrenocortical carcinoma is a lynch syndrome-associated cancer. J Clin Oncol. 2013;31(24):3012–8. Erratum in: J Clin Oncol. 2013;31(28):3612.
34. Mete O, Asa SL, Gill AJ, et al. Overview of the 2022 WHO classification of paragangliomas and pheochromocytomas. Endocr Pathol. 2022;33(1):90–114.
35. Thompson LD. Pheochromocytoma of the Adrenal gland Scaled Score (PASS) to separate benign from malignant neoplasms: a clinicopathologic and immunophenotypic study of 100 cases. Am J Surg Pathol. 2002;26(5):551–66.
36. Stenman A, Zedenius J, Juhlin CC. The value of histological algorithms to predict the malignancy potential of pheochromocytomas and abdominal paragangliomas—A meta-analysis and systematic review of the literature. Cancers (Basel). 2019;11(2):225.
37. Wu D, Tischler AS, Lloyd RV, et al. Observer variation in the application of the Pheochromocytoma of the Adrenal Gland Scaled Score. Am J Surg Pathol. 2009;33(4):599–608.
38. Kimura N, Takayanagi R, Takizawa N, et al. Pathological grading for predicting metastasis in phaeochromocytoma and paraganglioma. Endocr Relat Cancer. 2014;21(3):405–14.
39. Pierre C, Agopiantz M, Brunaud L, et al. COPPS, a composite score integrating pathological features, PS100 and SDHB losses, predicts the risk of metastasis and progression-free survival in pheochromocytomas/paragangliomas. Virchows Arch. 2019;474(6):721–34.
40. Stenman A, Zedenius J, Juhlin CC. Retrospective application of the pathologic tumor-node-metastasis classification system for pheochromocytoma and abdominal paraganglioma in a well characterized cohort with long-term follow-up. Surgery. 2019;166(5):901–6.
41. Juhlin CC. Challenges in paragangliomas and pheochromocytomas: from histology to molecular immunohistochemistry. Endocr Pathol. 2021;32(2):228–44.

42. Birtolo MF, Grossrubatscher EM, Antonini S, et al. Preoperative management of patients with ectopic Cushing's syndrome caused by ACTH-secreting pheochromocytoma: a case series and review of the literature. J Endocrinol Investig. 2023;46(10):1983–94.
43. Wang LL, Wei XJ, Zhang QC, Li F. Morphological and immunohistochemical characteristics associated with metastatic and recurrent progression in pheochromocytoma/paraganglioma: a cohort study. Ann Diagn Pathol. 2022;60:151981.
44. Juhlin CC, Mete O. Advances in adrenal and extra-adrenal paraganglioma: practical synopsis for pathologists. Adv Anat Pathol. 2023;30(1):47–57.

Medical Treatment in Advanced Adrenocortical Carcinoma

15

Valentina Cremaschi, Antonella Turla, Marta Laganà, and Deborah Cosentini

15.1 Introduction

Mitotane is the main component of standard systemic therapy. It is recommended by the international guidelines [1] and can be administered either as monotherapy or in combination with cisplatin, doxorubicin, and etoposide (EDP-M regimen). The efficacy of systemic therapy for advanced ACC is limited, but about 15% of patients survive at 5 years [1, 2] and about 2% may be disease-free for more than 5 years [3], indicating a potential, albeit limited, curative role. In this chapter we will provide an overview of the efficacy of standard therapy and suggest strategies for its optimal use. Additionally, we will present and briefly discuss the available data on targeted therapies and immunotherapies.

15.2 Standard Systemic Therapy: Mitotane

Mitotane is the only pharmacological compound approved for ACC, both in the adjuvant and in the advanced setting [1]. The drug is a dichlorodiphenyltrichloroethane (DDT) derivative and has a cytolytic effect on ACC cells and an inhibitory effect on adrenal steroidogenesis. Indeed, to avoid adrenocortical insufficiency, patients receiving mitotane need steroid replacement therapy. To maximize the drug's efficacy and tolerability, it is essential that its blood concentrations reach and maintain a therapeutic range of 14–20 mg/L. In the adjuvant setting, mitotane is indicated when the perceived risk of recurrence is high: stage III–IV and/or R1 and/or Ki67 >10%. If tolerated, the treatment should last for at least 2 years but no more

V. Cremaschi (✉) · A. Turla · M. Laganà · D. Cosentini
Medical Oncology, Department of Medical and Surgical Specialties, Radiological Sciences, and Public Health, University of Brescia at ASST Spedali Civili di Brescia, Brescia, Italy
e-mail: v.cremaschi@unibs.it; a.turla@unibs.it; marta.lagana@unibs.it; deborah.cosentini@unibs.it

© The Author(s) 2025
G. A. M. Tiberio (ed.), *Primary Adrenal Malignancies*, Updates in Surgery, https://doi.org/10.1007/978-3-031-62301-1_15

than 5 years. In the advanced setting, mitotane is administered in monotherapy for indolent and oligometastatic disease [3]. It can be combined with all the different locoregional treatments [1].

A key factor for the success of mitotane is the patient's compliance and active collaboration. In fact, this is a long-term treatment which may produce several side effects. For this reason, thorough counseling with clear explanations is mandatory, as is close medical monitoring and dose tailoring, as shown in Fig. 15.1.

15.3   Combination Therapy: Chemotherapy plus Mitotane (EDP-M)

The combination of mitotane with etoposide, doxorubicin and cisplatin (EDP-M regimen) is recommended [1, 4] in the case of:

– rapidly progressing disease;
– high burden of disease, with metastases in different organs;
– disease progressing during mitotane monotherapy, with mitotane blood levels in the therapeutic range, which means that mitotane alone is not enough to control the tumor growth.

The EDP regimen, which is usually administered for a maximum of 6–8 cycles, combines three effective cytotoxic compounds:

– etoposide (or VP-16) that inhibits topoisomerase II, whose activity prevents DNA unwinding;
– doxorubicin that inhibits DNA synthesis and transcription;
– cisplatin that inhibits DNA synthesis and function as well as its transcription.

The recommendation for the use of EDP-M is supported by the FIRM-ACT trial, a multi-center prospective international study which enrolled 304 patients and compared the efficacy of the EDP-M regimen to the streptozotocin-mitotane regimen. Patients in the EDP-M arm had better median progression-free survival (mPFS) (5.0 months vs. 2.1 months; hazard ratio 0.55, 95% CI 0.43–0.69, $p < 0.001$) and longer median overall survival (mOS), which just failed to attain statistical significance (14.8 months versus 12.0 months; $p = 0.07$) [2]. In a retrospective trial performed at our Institution, we analyzed the data of 58 patients affected by advanced/metastatic ACC, who received a median of 5.5 EDP cycles. According to the Response Evaluation Criteria in Solid Tumors (RECIST) 1.1 criteria, no complete response was achieved. Fifty percent of patients obtained a partial response and in 26% of cases a stability of disease was documented; mPFS and mOS were 10.1 (95% CI 8.1–12.8) and 18.7 (95% CI 14.6–22.8) months, respectively [5].

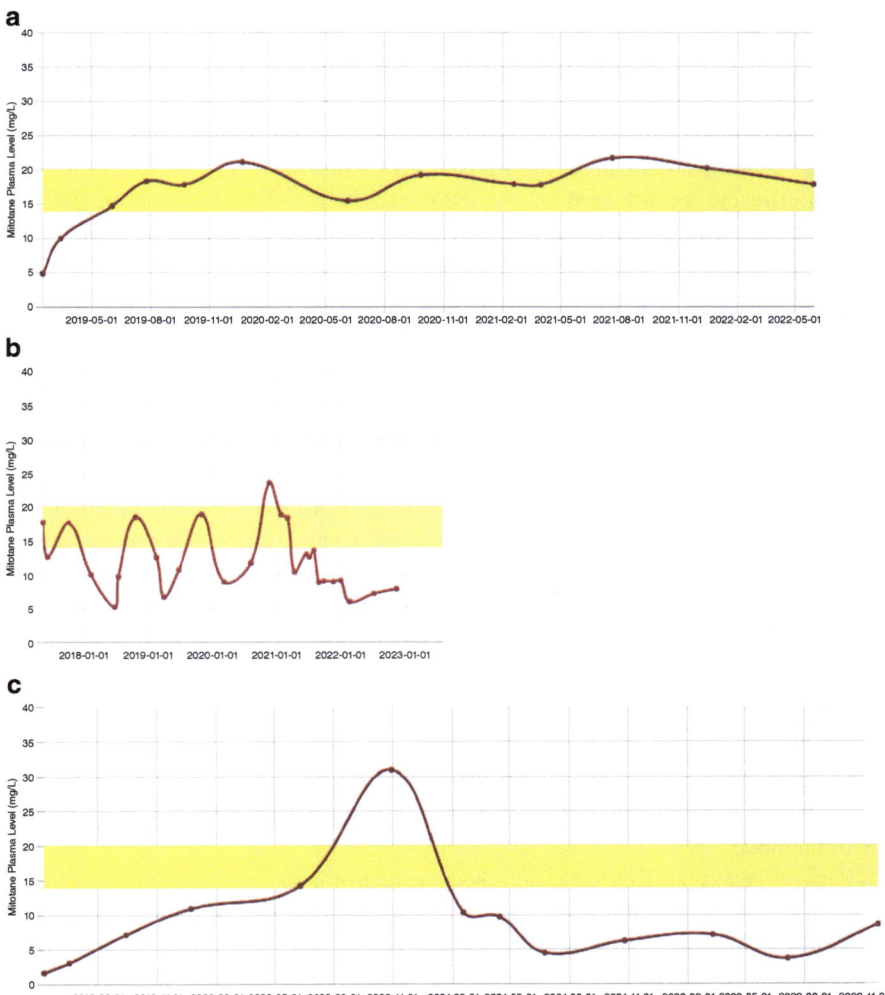

Fig. 15.1 Three examples of variation in serum mitotane levels. The first case (**a**) describes the optimal situation, in which the serum concentration is reached and maintained in the therapeutic range. The second case (**b**) describes a patient who temporarily stopped the drug due to toxicity, causing the serum drug concentration to be inhomogeneous and well below the therapeutic range. The third case (**c**) describes a patient with poor compliance who reached the therapeutic range but spontaneously reduced the drug owing to the appearance of gynecomastia. All three patients with metastatic disease received mitotane for a long period of time. Noteworthy, in the last case, the patient was persuaded to increase the drug dosage after experiencing disease progression

EDP-M therapy has the potential to be very effective in a minority of patients, leading to a limited number of disease responses [2]. For this reason, this chemotherapy regimen needs to be administered in the most efficient way. It is important to treat advanced ACC patients in oncologic referral centers with dedicated multidisciplinary teams and extensive experience in the treatment of ACC.

When monitoring a patient during EDP-M treatment, it should be kept in mind that its efficacy is influenced by the cytotoxic effects of both chemotherapy and mitotane, the latter being notoriously delayed. It is also important to note that an early radiological size increase EDP-M treatment does not necessarily correspond to disease progression. A multidisciplinary evaluation is necessary to assess the response to EDP-M, considering multiple parameters such as RECIST 1.1, the Choi criteria, and tumor volume [6]. In any case, EDP-M should not be withdrawn in cases of early progression if the mitotane blood level is below the therapeutic concentrations, unless new lesions appear [5].

The EDP-M scheme can be burdened by significant toxicities and their correct management is recommended [7]. Neutropenia occurs in 77% of patients at nadir and in 53% of patients at recycling. Nevertheless, EDP efficacy depends on the correct administration of full doses. Granulocyte colony-stimulating factors (G-CSF) should be administered 24–48 h after every cycle. The presence of Cushing's syndrome is the sole factor which may prevent the administration of EDP at full doses, due to the increased risk of infections and sepsis. Nausea and vomiting affect up to 90% of patients during chemotherapy administration and in the following days [7]. Asthenia is also a frequent symptom (70% of cases). The presence of mitotane-induced hypoadrenalism should be considered if patients experience nausea, vomiting and asthenia after EDP-M, since distinguishing between chemotherapy-induced nausea/asthenia and hypoadrenalism can be challenging [4, 7]. In cases of uncertainty, an extra dose of glucocorticoids is recommended.

Patients receiving EDP-M could develop neurological toxicity due both to cisplatin (peripheral) and mitotane (central). Since the mechanism of nerve damage is different, mitotane should not be withdrawn in the case of cisplatin-related neurotoxicity [7].

Furthermore, an ongoing trial (ACACIA, NCT03723941) is evaluating the efficacy of the addition of etoposide and cisplatin to mitotane therapy (EP-M regimen) in the adjuvant setting in resected patients at high risk of relapse.

15.4   Beyond EDP-M

The second-line treatments tested for patients who progress on EDP-M have not produced sufficient results to be considered standard therapy. The combination of gemcitabine plus capecitabine [8] or streptozocin [2] led to a limited response rate and poor PFS (between 2 and 4 months). Cabazitaxel was found to be totally inactive in a prospective phase II trial [9] and these results were in contrast with a previous in vitro experiment. Even though temozolomide was highly active *in vitro*, it obtained an objective response rate (ORR) of 36% and a mPFS and mOS of only 3.5 and 7.2 months, respectively [10].

Several trials have tested targeted drugs in advanced ACC. Results with anti-epidermal growth factor receptors (EGFR) and anti-angiogenic drugs, either alone or in combination with chemotherapy, have been disappointing [11]. Three cases of objective response and three of stable disease in 16 pretreated ACC patients described with cabozantinib, a tyrosine kinase inhibitor (TKI) targeting c-Met, vascular endothelial growth factor receptor-2 (VEGFR-2), AXL, and RET, are of some interest and deserve confirmation [12]. Dovitinib, a targeted therapy that inhibits the fibroblast growth factor receptor (FGFR), reached no objective response, but a clinical benefit longer than 6 months was achieved in 35% of patients, with long-lasting stable disease in 23% [13]. Moreover, thalidomide, with its anti-inflammatory and anti-angiogenetic properties, yielded poor results in a retrospective study: 7.5% stable disease, and no radiological response according to the RECIST criteria [14].

The overall disappointing results of these molecular target drugs can be partly explained by the fact that none of them interact with the three molecular pathways that are the basis of ACC pathophysiology: insulin-like growth factor, p53 and WNT/β-catenin. As regards the insulin-like growth factor pathways IGF1-IGF2, the IGF1 receptor inhibitor cixutumumab, combined with mitotane as a first-line treatment for advanced/metastatic ACC, led to a mPFS of 6 weeks (range 2.66–48) and a clinical benefit in 8 out of 20 patients (1 partial response, 7 stable disease) [15]. In a trial exploring cixutumumab plus the mTOR (mammalian target of rapamycin) inhibitor temsirolimus, 11 of 26 patients (42%) achieved stable disease for more than 6 months [16]. Linsitinib, an oral inhibitor of both the IGF1 receptor and the insulin receptor, failed to demonstrate any superiority over placebo in terms of both mPFS and mOS in heavily pretreated ACC patients [17].

Theragnostics is another modern therapeutic strategy in the treatment of cancer patients. Radionuclide therapy has also been tested in patients with ACC in small clinical trials. More than 50% of ACC cells have been found to express somatostatin receptors (SSTRs). A study [18] conducted on 19 pretreated metastatic ACC patients showed radiometabolic uptake of any intensity in 8 (42%) patients and a strong uptake in 2 (11%) patients; both patients were treated with ^{177}Lu-DOTATATE obtaining a disease control of 4 and 12 months. The radionuclide molecule ^{131}I-IMAZA, which targets 11-beta hydroxylase, has also been tested for ACC treatment [19]. Thirteen patients underwent a median dose of 25.7 GBq ^{131}I-IMAZA (range 18.1–30.7 GBq) and follow-up data were available for 12 patients. Stable disease was obtained in five of them with a mPFS of 14.3 months (range 8.3–21.9). The mOS in the intention-to-treat population was 14.1 months (4.0–56.5). These results suggest that the theragnostic approach should continue to be tested in patients with ACC.

ACC has an intrinsic immunoresistance because of its pathophysiological pathways (β-catenin gene activation and *TP53* mutation), which are also mechanisms of resistance to immunotherapy, and because of the frequent glucocorticoid hypersecretion which generates an immunosuppressive microenvironment. Despite ACC immunoresistance, many clinical trials testing PD1 (programmed death protein 1),

PDL-1 (programmed death ligand-1) and CTLA-4 (cytotoxic T-lymphocyte antigen-4) inhibitors have been performed with interesting results, which appear more promising compared with those obtained with targeted molecules [11]. In a phase II trial, pembrolizumab obtained an ORR of 23%, with a mPFS of 2.1 months and a mOS of nearly 25 months. Nivolumab was tested in a phase II trial and an ORR of 10%, a mPFS of nearly 2 months and a mOS of 21 months were achieved. In a trial evaluating avelumab, the ORR was 6%, and the mPFS and mOS were 2.6 and 10.6 months, respectively. The CTLA-4 inhibitor ipilimumab was tested in combination with nivolumab in two phase II trials and the ORR was 6% and 33% [11, 20].

Many immunotherapy trials are currently enrolling patients with metastatic ACC, and this pharmacological approach could be one of the most important in the treatment of ACC.

References

1. Fassnacht M, Assie G, Baudin E, et al. Adrenocortical carcinomas and malignant phaeochromocytomas: ESMO-EURACAN Clinical Practice Guidelines for diagnosis, treatment and follow-up. Ann Oncol. 2020;31(11):1476–90. Erratum in: Ann Oncol. 2023;34(7):631.
2. Fassnacht M, Terzolo M, Allolio B, et al. Combination chemotherapy in advanced adrenocortical carcinoma. N Engl J Med. 2012;366(23):2189–97.
3. Terzolo M, Daffara F, Ardito A, et al. Management of adrenal cancer: a 2013 update. J Endocrinol Invest. 2014;37(3):207–17.
4. Berruti A, Terzolo M, Sperone P, et al. Etoposide, doxorubicin and cisplatin plus mitotane in the treatment of advanced adrenocortical carcinoma: a large prospective phase II trial. Endocr Relat Cancer. 2005;12(3):657–66.
5. Laganà M, Grisanti S, Cosentini D, et al. Efficacy of the EDP-M scheme plus adjunctive surgery in the management of patients with advanced adrenocortical carcinoma: the Brescia experience. Cancers (Basel). 2020;12(4):941.
6. Ambrosini R, Balli MC, Laganà M, et al. Adrenocortical carcinoma and CT assessment of therapy response: the value of combining multiple criteria. Cancers (Basel). 2020;12(6):1395.
7. Turla A, Laganà M, Grisanti S, et al. Supportive therapies in patients with advanced adrenocortical carcinoma submitted to standard EDP-M regimen. Endocrine. 2022;77(3):438–43.
8. Henning JEK, Deutschbein T, Altieri B, et al. Gemcitabine-based chemotherapy in adrenocortical carcinoma: a multicenter study of efficacy and predictive factors. J Clin Endocrinol Metab. 2017;102(11):4323–32.
9. Laganà M, Grisanti S, Ambrosini R, et al. Phase II study of cabazitaxel as second-third line treatment in patients with metastatic adrenocortical carcinoma. ESMO Open. 2022;7(2):100422.
10. Cosentini D, Badalamenti G, Grisanti S, et al. Activity and safety of temozolomide in advanced adrenocortical carcinoma patients. Eur J Endocrinol. 2019;181(6):681–9.
11. Cremaschi V, Abate A, Cosentini D, et al. Advances in adrenocortical carcinoma pharmacotherapy: what is the current state of the art? Expert Opin Pharmacother. 2022;23(12):1413–24.
12. Kroiss M, Megerle F, Kurlbaum M, et al. Objective response and prolonged disease control of advanced adrenocortical carcinoma with cabozantinib. J Clin Endocrinol Metab. 2020;105(5):1461–8.
13. García-Donas J, Hernando Polo S, Guix M, et al. Phase II study of dovitinib in first line metastatic or (non resectable primary) adrenocortical carcinoma (ACC): SOGUG study 2011-03. J Clin Oncol. 2014;32(15 suppl):4588.
14. Kroiss M, Deutschbein T, Schlötelburg W, et al. Treatment of refractory adrenocortical carcinoma with thalidomide: analysis of 27 patients from the European Network for the Study of Adrenal Tumours Registry. Exp Clin Endocrinol Diabetes. 2019;127(9):578–84.

15. Lerario AM, Worden FP, Ramm CA, et al. The combination of insulin-like growth factor receptor 1 (IGF1R) antibody cixutumumab and mitotane as a first-line therapy for patients with recurrent/metastatic adrenocortical carcinoma: a multi-institutional NCI-sponsored trial. Horm Cancer. 2014;5(4):232–9. Erratum in: Horm Cancer. 2014;5(6):424.
16. Naing A, Lorusso P, Fu S, et al. Insulin growth factor receptor (IGF-1R) antibody cixutumumab combined with the mTOR inhibitor temsirolimus in patients with metastatic adrenocortical carcinoma. Br J Cancer. 2013;108(4):826–30.
17. Fassnacht M, Berruti A, Baudin E, et al. Linsitinib (OSI-906) versus placebo for patients with locally advanced or metastatic adrenocortical carcinoma: a double-blind, randomised, phase 3 study. Lancet Oncol. 2015;16(4):426–35.
18. Grisanti S, Filice A, Basile V, et al. Treatment with 90Y/177Lu-DOTATOC in patients with metastatic adrenocortical carcinoma expressing somatostatin receptors. J Clin Endocrinol Metab. 2020;105(3):dgz091.
19. Hahner S, Hartrampf PE, Mihatsch PW, et al. Targeting 11-beta hydroxylase with [131I]IMAZA: a novel approach for the treatment of advanced adrenocortical carcinoma. J Clin Endocrinol Metab. 2022;107(4):e1348–55.
20. Pegna GJ, Roper N, Kaplan RN, et al. The immunotherapy landscape in adrenocortical cancer. Cancers (Basel). 2021;13(11):2660.

Integrated Approach in Locally Advanced, Oligometastatic or Recurrent Adrenocortical Carcinoma

16

Antonella Turla, Deborah Cosentini, Alfredo Berruti, and Guido A. M. Tiberio

16.1 Introduction

Radical surgery represents the only possibility to cure adrenocortical carcinoma (ACC). Unfortunately, ACC is frequently diagnosed at an advanced stage, in the form of locally advanced or metastatic disease. Furthermore, a significant proportion of patients experience recurrent or metastatic disease after surgery performed with curative intent. Although re-do resection may offer good survival results in these cases, the occurrence of "surgical cure" is anecdotal. In recent years, multimodal approaches integrating systemic and local treatments have emerged, widening the spectrum of integrated treatments for advanced stage, recurrent or oligometastatic patients. We will discuss these approaches in the following pages, well aware that future improvements and the broadening of treatment options will originate from multidisciplinary collaboration.

16.2 Neoadjuvant Chemotherapy Followed by Surgery

Surgery is the cornerstone of therapy and the only treatment modality which may offer a chance of cure; it should be considered in the treatment of all patients and at any stage of disease with the aim of removing the entire tumor bulk (R0). However, while upfront surgery is a clear option in early stages, it may not be the best strategy

A. Turla · D. Cosentini · A. Berruti (✉)
Medical Oncology, Department of Medical and Surgical Specialties, Radiological Sciences, and Public Health, University of Brescia at ASST Spedali Civili di Brescia, Brescia, Italy
e-mail: a.turla@unibs.it; deborah.cosentini@unibs.it; alfredo.berruti@unibs.it

G. A. M. Tiberio
General Surgery, Department of Clinical and Experimental Sciences, University of Brescia at ASST Spedali Civili di Brescia, Brescia, Italy
e-mail: guido.tiberio@unibs.it

© The Author(s) 2025
G. A. M. Tiberio (ed.), *Primary Adrenal Malignancies*, Updates in Surgery,
https://doi.org/10.1007/978-3-031-62301-1_16

for patients with locally advanced or metastatic disease, as in these instances the likelihood of recurrence is high. Considering that an objective response can be achieved in 50% of cases using the standard systemic EDP-M scheme (etoposide, doxorubicin, cisplatin plus mitotane) [1, 2], in patients with surgically amenable ENSAT stage III–IV ACC [3] and in those with borderline resectable (BR) ACC the possibility of neoadjuvant chemotherapy followed by surgery should be carefully discussed by the multidisciplinary team.

Neoadjuvant therapy for ACC is a fairly new practice founded on strong theoretical principles. These can be summarized as follows: to reduce tumor size and thus increase the likelihood of R0 resection increasing at the same time the possibility to preserve other organs; to administer nephrotoxic systemic therapy before ipsilateral nephrectomy (if indicated); to select for radical surgery those responding patients who are most likely to achieve the best outcome. In the case of locally advanced disease where radical surgery is not feasible, neoadjuvant chemotherapy is the only option, but this treatment modality should also be considered in patients with BR disease.

BR disease is a multifold definition including anatomical, biological and patient-related criteria. As for the anatomical definition, this applies to patients requiring multi-organ or vascular resections and, in general, those at high risk for a margin-positive resection based on preoperative imaging. The biological definition identifies those patients with suspicion of metastases or potentially resectable oligometastatic disease. Finally, the patient-related definition applies to patients with significant comorbidities contraindicating upfront surgery; among these, poor patient condition, pulmonary embolism and metabolic failure secondary to hormone hyperproduction are the acute conditions more frequently observed.

Some evidence supports the combination of systemic therapy and surgery. A retrospective study analyzed neoadjuvant chemotherapy (mainly EDP-M) in BR ACC [4]. Fifty-three individuals operated on with curative intent were included in the study; of them, 15 (28%) were classified as BR and received neoadjuvant therapy. Despite a more advanced clinical stage at diagnosis and a higher incidence of multi-visceral resections, the rate of margin-positive resections was similar in BR and resectable patients; furthermore, the pathologic size of the tumor was also similar in the study groups, despite larger tumor diameters at diagnosis in the BR group. Median disease-free survival (DFS) for individuals with resected BR ACC was 28 months (95% CI 2.9–not achieved) as opposed to 13 months (95% CI 5.8 to 46.9; p = 0.40) for individuals with resectable disease. Five-year overall survival (OS) showed no significant difference: 65% for the neoadjuvant group (N = 13) versus 50% for the upfront surgery patients (N = 38) (p = 0.72).

Another retrospective Italian study explored the sequence of EDP-M and surgical resection in ACC diagnosed as locally advanced and non-resectable (6 cases) or metastatic [2]. The median progression-free survival (PFS) and OS for the cohort of 58 patients were 10.1 months and 18.7 months, respectively. Surgery was indicated in 26 (45%) responding patients to remove the remaining disease or reduce the tumor bulk: 13 attained a disease-free status and 13 had a residual disease ≤10% (R1 resection in 5 and small lung metastases in 8). Post-chemotherapy surgery provided some important information: the possibility to identify complete responders to chemotherapy (7%) and to reassess Ki67 expression in post-chemotherapy tumor specimens. As expected, responding patients subjected to surgical resection showed better outcomes: median DFS and OS were 13.1 and 29.8 months, respectively.

In the absence of a randomized clinical study, from these papers we cannot conclude that the association of chemotherapy and surgery is *per se* more efficacious than surgery, since the patients operated on were selected among responders, who already benefit from survival advantages. Considering that the disease's biologic behavior is the main determinant of the long-term success of surgery, neoadjuvant chemotherapy is therefore helpful in selecting for surgery those patients with either an indolent primary or affected by a disease which has been made indolent by chemotherapy. Therefore, neoadjuvant therapy is a valid option for patients with BR or unresectable disease at diagnosis. Moreover, potentially resectable patients who demonstrate aggressive disease, such as those displaying a significant volume increase in two imaging procedures performed sequentially within a short period of time, or patients who have already undergone surgery but have developed an early recurrence, are also eligible. Considering these premises, it is necessary for a multidisciplinary team to carefully discuss any patient with ACC. The Brescia experience shows that an aggressive multimodal and staged treatment offers interesting results, permitting long-term disease-free status to be obtained in a non-negligible percentage of cases.

16.3 Adrenalectomy: Upfront or After Primary Chemotherapy in the Metastatic Setting?

Surgical excision of the primary tumor is effective in improving survival in some metastatic malignancies. This strategy may therefore be an option in patients with metastatic ACC. The role of non-curative surgical debulking was explored in a retrospective cohort of 239 metastatic patients by the American-Australian-Asian collaborative group [5]. A propensity score analysis using as matching criteria the patients' age and the number of metastatic organs (2 or >2) investigated OS in patients treated with or without resection of the adrenal primary. Patients in the surgery group had a median OS of 25.2 months (95% CI 21.0–29.5) as opposed to those in the no-resection group, whose median OS was 9.0 months (95% CI 6.7–11.3). From this analysis, age (HR [hazard ratio] 1.02; 95% CI 1.00–1.03), hormone excess (HR 2.56; 95% CI 1.66–3.92) and local treatment of metastasis (HR 0.41; 95% CI 0.47–0.65) also emerged as independent predictors of survival.

A multivariate analysis of 202 patients with synchronous metastatic ACC identified from the SEER (Surveillance, Epidemiology, and End Results) database [3], showed that those 76 (37.6%) patients who underwent adrenal surgery had better survival as compared to non-surgical patients (median OS: 13 vs 4 months, $p < 0.001$). Besides, adrenalectomy (HR 0.64; 95% CI 0.45–0.92; $p = 0.017$), metastasectomy (HR 0.48; 95% CI 0.26–0.86; P = 0.013) and chemotherapy (HR 0.59; 95% CI 0.42–0.82; p = 0.002) were also associated with improved survival.

Based on these data, surgical removal of the adrenal primary should always be considered in metastatic ACC, especially in the oligometastatic status. The critical point is whether to perform surgery upfront or after chemotherapy. The position of the multidisciplinary group in Brescia is that all patients should undergo an upfront systemic treatment and that the surgical indication should be discussed on the basis of a computed tomography (CT) re-evaluation after chemotherapy [6]. Generally, surgery should be offered only to those who attain disease response or stabilization. This approach has several advantages: (1) patients not progressing after systemic treatment are selected as having a more indolent disease either spontaneously or after systemic treatment, (2) surgery after chemotherapy has the opportunity to identify any complete pathological response which represents undisputed evidence of treatment efficacy and a powerful prognostic parameter, (3) surgery after chemotherapy provides the unique opportunity to reassess the tumor biology and obtain useful information for planning the next treatment. On this subject, the Brescia experience has shown that proliferation activity assessed by Ki67 in post-chemotherapy residual tumor has a stronger prognostic role than Ki67 evaluated under baseline conditions [2].

In metastatic patients, the surgical treatment of metastases should be considered. Metastasectomy in the synchronous setting should be indicated in selected cases, at completion of removal of the primary, if R0 status is anticipated on preoperative cross-sectional imaging and in the case of favorable biological behavior [7–9]. The same criteria should also be adopted, in association with mitotane, in patients with metachronous metastatic disease [10, 11].

16.4 Cytoreduction and Hyperthermic Intra-peritoneal Chemotherapy

In the major referral centers, cytoreductive surgery completed by hyperthermic intraperitoneal chemotherapy (HIPEC) is now part of the toolbox for the multi-modal management of different primaries with peritoneal involvement. This strategy pursues the optimization of local control through an aggressive surgical strategy aimed to maximal cytoreduction of the neoplastic bulk and the synergistic effect of chemotherapy and hyperthermia on low-volume (≤ 2 mm) or microscopic neoplastic remnants.

As far as ACC is concerned, the literature is scarce, consisting of a few case reports, and the first paper reporting on the value of HIPEC was published by Hughes et al. in 2018 [12]. In a selected series of 10 patients who developed local and/or peritoneal ACC recurrence at least 12 months after adrenalectomy a complete cytoreduction (CCR = 0) was achieved. HIPEC was carried out using cisplatin (250 mg/m^2/L of perfusate) at 40 °C for 90 min. The complication rate was 40%

with complications graded 2 or 3 according to the Clavien-Dindo scale. At a median follow-up of 23 months, median DFS was 19 months while median OS was not reached.

A second paper was published by our group in 2020 [13]. HIPEC was conducted in a different way, using cisplatin (25 mg/m^2/L) and doxorubicin (4.5 mg/m^2/L) at 42 °C for 60 min. We analyzed a small cohort of 14 patients presenting local or peritoneal recurrence of ACC, responsive or stable after chemotherapy (EDP-M or mitotane alone). The morbidity rate reached 77%, due to extended (often multi-organ) resections required by the recurrent neoplastic bulk; grade 3 and 4 complications were observed in 12% and 6% of cases, respectively. Mortality was nil. After a median follow-up of 30 months, patients with recurrence showed a median local/peritoneal DFS of 12 months while median OS was not reached. Interestingly and for the first time, in the same paper, we also presented data concerning a cohort of 13 BR ACC patients subjected to cytoreductive surgery and HIPEC with prophylactic intent. In this setting, the 90-day morbidity rate was 46% with grade 3 and 4 complications accounting for 15% and 8%, respectively. After a median follow-up of 25 months, both median DFS and OS was not attained.

For this monograph, we revised our database. Twenty-three patients received 26 cytoreductive procedures and HIPEC for the management of recurrent ACC. All but one patient received chemotherapy (EDP-M or mitotane alone) after the diagnosis of recurrence. Twenty percent of our patients were alive and disease-free 5 years after the procedure. We were also able to confirm the results reported by Hughes et al.: in those patients (n = 10) who recurred ≥12 months after adrenal surgery, median local/peritoneal DFS was 19 months and, after a median follow-up of 27 months, median OS was not attained. Considering the prophylactic setting, in a cohort of 26 patients presenting with stage II and III ACC, we made a comparison between a group of 11 patients subjected to regional adrenalectomy completed by HIPEC and a group of 15 patients subjected only to regional adrenalectomy. After a median follow-up of 25 and 34 months, respectively, the median local/peritoneal DFS and OS was not reached in either group. Albeit not significant, Kaplan-Meyer survival curves show an advantage of 20% in terms of local/peritoneal DFS at 2 years, which is replicated in the 3-year OS (Fig. 16.1).

16.4.1 Alternative Locoregional Treatments

Control of the tumor bulk by means of local treatments supports systemic therapy and exerts a paramount prognostic role in the management of metastatic ACC. The multidisciplinary team may choose among different options such as radiation therapy, radiofrequency or microwave ablation, cryoablation and transcatheter arterial chemoembolization. Unfortunately, due to the rarity of the disease, the specific literature is scarce. For this reason, besides general selection criteria, the choice of the technique is influenced by the experience of the local team. Interestingly, the

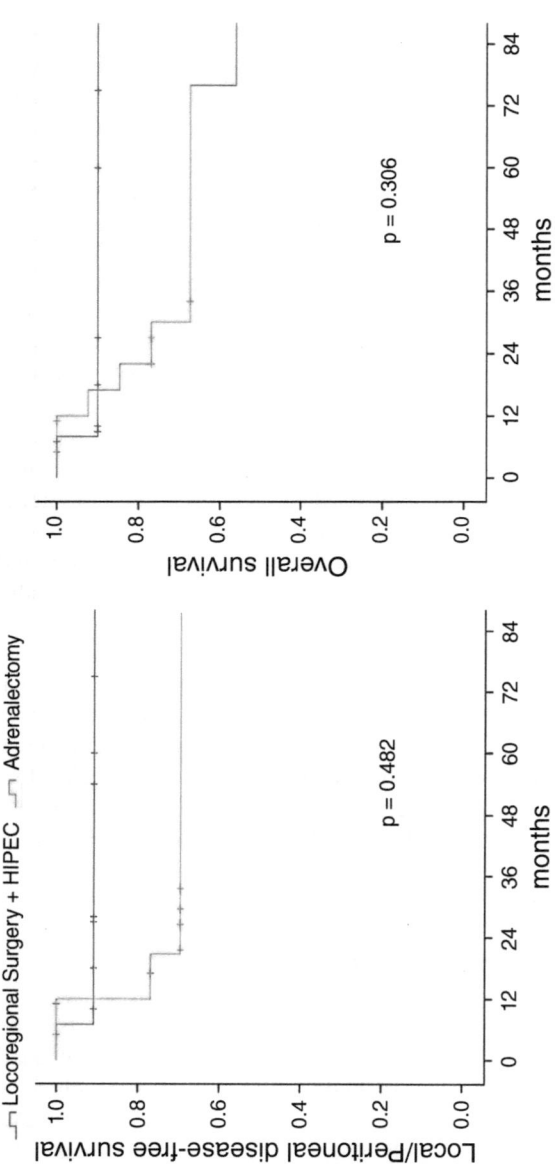

Fig. 16.1 Local/peritoneal disease-free survival and overall survival in stage II & III adrenocortical carcinoma. Effect of HIPEC (cytoreduction and hyperthermic intraperitoneal chemotherapy) at completion of regional surgery

combination of locoregional treatment and mitotane allowed, in a selected case series, the start of chemotherapy to be significantly delayed in patients with metachronous oligometastatic disease [14].

Small series studied the ablative treatment of hepatic metastases employing radiofrequency or microwave ablation [15, 16]. These techniques show particular value in cases of oligometastatic disease and diameter of the metastases ≤3 cm, in particular if employed in combination with systemic treatments [17]. Transcatheter arterial chemoembolization (TACE) has also been reported in the management of metastatic ACC. As for other primaries, it may be instrumental in controlling or stabilizing hepatic metastases. Cisplatin, doxorubicin, and/or mitomycin or lipiodol alone have been selectively injected in the arterial branches directed to the metastatic lesions [10].

References

1. Fassnacht M, Terzolo M, Allolio B, et al. Combination chemotherapy in advanced adrenocortical carcinoma. N Engl J Med. 2012;366(23):2189–97.
2. Laganà M, Grisanti S, Cosentini D, et al. Efficacy of the EDP-M scheme plus adjunctive surgery in the management of patients with advanced adrenocortical carcinoma: the Brescia experience. Cancers (Basel). 2020;12(4):941.
3. Wu K, Liu Z, Li X, Lu Y. Adrenal surgery for synchronously metastatic adrenocortical carcinoma: a population-based analysis. World J Surg. 2021;45(5):1457–65.
4. Bednarski BK, Habra MA, Phan A, et al. Borderline resectable adrenal cortical carcinoma: a potential role for preoperative chemotherapy. World J Surg. 2014;38(6):1318–27.
5. Srougi V, Bancos I, Daher M, et al. Cytoreductive surgery of the primary tumor in metastatic adrenocortical carcinoma: impact on patients' survival. J Clin Endocrinol Metab. 2022;107(4):964–71.
6. Cosentini D, Laganà M, Turla A, et al. Letter to the Editor from Cosentini et al: "Cytoreductive surgery of the primary tumor in metastatic adrenocortical carcinoma: impact on patients' survival". J Clin Endocrinol Metab. 2022;107(7):e3092–3.
7. Ettaieb MH, Duker JC, Feelders RA, et al. Synchronous vs. metachronous metastases in adrenocortical carcinoma: an analysis of the Dutch Adrenal Network. Horm Cancer. 2016;7(5–6):336–44.
8. Prendergast KM, Smith PM, Tran TB, et al. Features of synchronous versus metachronous metastasectomy in adrenal cortical carcinoma: analysis from the US adrenocortical carcinoma database. Surgery. 2020;167(2):352–7.
9. Dy BM, Strajina V, Cayo AK, et al. Surgical resection of synchronously metastatic adrenocortical cancer. Ann Surg Oncol. 2015;22(1):146–51.
10. op den Winkel J, Pfannschmidt J, Muley T, et al. Metastatic adrenocortical carcinoma: results of 56 pulmonary metastasectomies in 24 patients. Ann Thorac Surg. 2011;92(6):1965–70.
11. Gaujoux S, Al-Ahmadie H, Allen PJ, et al. Resection of adrenocortical carcinoma liver metastasis: is it justified? Ann Surg Oncol. 2012;19(8):2643–51.
12. Hughes MS, Lo WM, Beresnev T, et al. A phase II trial of cytoreduction and hyperthermic intraperitoneal chemotherapy for recurrent adrenocortical carcinoma. J Surg Res. 2018;232:383–8.
13. Tiberio GAM, Ferrari V, Ballarini Z, et al. Hyperthermic intraperitoneal chemotherapy for primary or recurrent adrenocortical carcinoma. A single center study. Cancers (Basel). 2020;12(4):969.
14. Roux C, Boileve A, Faron M, et al. Loco-regional therapies in oligometastatic adrenocortical carcinoma. Cancers (Basel). 2022;14(11):2730.

15. Wood BJ, Abraham J, Hvizda JL, et al. Radiofrequency ablation of adrenal tumors and adrenocortical carcinoma metastases. Cancer. 2003;97(3):554–60.
16. Li X, Fan W, Zhang L, et al. CT-guided percutaneous microwave ablation of adrenal malignant carcinoma: preliminary results. Cancer. 2011;117(22):5182–8.
17. Megerle F, Kroiss M, Hahner S, Fassnacht M. Advanced adrenocortical carcinoma—What to do when first-line therapy fails? Exp Clin Endocrinol Diabetes. 2019;127(2–03):109–16.

Medical Treatment of Malignant Pheochromocytoma

17

Marta Laganà, Deborah Cosentini, Antonella Turla, Valentina Cremaschi, Salvatore Grisanti, and Alfredo Berruti

17.1 Chemotherapy

Systemic cytotoxic chemotherapy is the oldest available treatment for progressive metastatic pheochromocytoma/paraganglioma (mPPGL) and may be the only treatment available in many countries around the world [1, 2]. Chemotherapy can be effective in some patients and the best-studied protocol includes a combination of cyclophosphamide, vincristine, and dacarbazine (CVD), sometimes with the addition of doxorubicin. A meta-analysis and systematic review of the literature evaluating the best studies published on CVD chemotherapy in mPPGL through 2014 indicated that approximately 37% of patients benefit from chemotherapy. However, this number could be an overestimation, because only one of the studies clearly included patients with disease progression [3]. Nevertheless, chemotherapy may stop tumor progression, decrease tumor size, decrease hormonal secretion and improve symptoms of catecholamine excess, prevent complications related to the tumor location and burden (e.g., skeletal events), and perhaps improve overall survival. Complete responses to CVD chemotherapy are exceptional [2–6].

CVD toxicity is mainly characterized by fatigue, bone marrow suppression, peripheral neuropathy, nausea, vomiting, and constipation. Patients with hormonally active tumors must be carefully treated with alpha- and beta-blockers because CVD chemotherapy may lead to tumor-cell destruction with a subsequent release of catecholamines, leading to a hypertensive crisis. In addition, patients must receive counseling on maintaining a diet rich in fiber and using laxatives to prevent or treat gastrointestinal dysmotility; of note, constipation makes it more difficult to tolerate chemotherapy, especially in patients with noradrenaline-secreting tumors.

M. Laganà (✉) · D. Cosentini · A. Turla · V. Cremaschi · S. Grisanti · A. Berruti
Medical Oncology, Department of Medical and Surgical Specialties, Radiological Sciences, and Public Health, University of Brescia at ASST Spedali Civili di Brescia, Brescia, Italy
e-mail: marta.lagana@unibs.it; deborah.cosentini@unibs.it; a.turla@unibs.it; v.cremaschi@unibs.it; salvatore.grisanti@unibs.it; alfredo.berruti@unibs.it

© The Author(s) 2025
G. A. M. Tiberio (ed.), *Primary Adrenal Malignancies*, Updates in Surgery,
https://doi.org/10.1007/978-3-031-62301-1_17

Occasionally, a patient may develop severe, even lethal constipation [7]. Nausea and vomiting happen frequently; antiemetics are recommended. The use of metoclopramide is contraindicated in hormonally active tumors [8].

CVD chemotherapy is mainly recommended for tumors characterized by rapid progression. In addition, the North American Neuroendocrine Tumor Society guidelines [3] recommend considering the use of chemotherapy for patients with bulky disease and intense symptoms related to tumor burden (e.g., pain) or hormonal excess. There is no standard definition of rapid progression; clinical assessment and experience determine when to recommend chemotherapy. The mean number of cycles required to achieve an oncologic benefit is still to be determined. In our institution, patients are treated for a period of 4–12 months, with radiographic evaluations every 2–3 months. The duration of treatment depends on the tumor response, the patient's ability to tolerate the side effects, and the potential risk for bone marrow dysplasia and leukemia. Patients who respond to CVD chemotherapy are frequently transitioned to a maintenance regimen of temozolomide. The side effects of temozolomide are milder than those of CVD.

Both dacarbazine and temozolomide are alkylating agents, and chronic use of these drugs is associated with a small but cumulative risk of myelodysplasias and leukemias. Previous studies of temozolomide have limited the total duration of treatment to 12 months. Because therapeutic options are limited, the prolonged use of CVD or temozolomide depends on the availability of other therapeutic options, oncologic benefits, and tolerability of side effects; therefore, these treatments may be prescribed for longer than 12 months on an individual basis. Temozolomide as a first-line treatment for mPPGL has also been described, and this treatment has been associated with oncologic and biochemical responses in very small studies and case reports [9, 10]. A randomized, prospective clinical phase II trial of temozolomide alone compared with temozolomide plus olaparib (ALLIANCE A021804 trial; ClinicalTrials.gov Identifier: NCT04394858) will provide additional guidance on the use of temozolomide in clinical practice.

Currently, we do not have predictive factors for CVD chemotherapy or temozolomide response. The largest study on CVD chemotherapy indicated that the larger the tumor burden, the less effective chemotherapy is. A few studies have suggested that the presence of *SDHB* mutations predicts a response to CVD chemotherapy or temozolomide [11].

However, these studies are not prospective and lack a comparative group. In addition, the largest study on chemotherapy showed that some *SDHB* carriers did not respond to treatment, which is concordant with the clinical experiences of many referral centers [2].

17.2   Targeted Therapy

Angiogenesis, tumor proliferation, tumor invasion, and the development of metastases are important hallmarks of cancer which are regulated by different tyrosine kinase receptors. These receptors could be inhibited by small tyrosine kinase inhibitors (TKIs) that are currently approved for the treatment of many different malignancies [12]. Most of these molecules target several receptors and some are under evaluation in clinical trials for mPPGL. Preliminary results indicate that TKIs may cause rapid tumor size reduction, disease stabilization, and improvement of symptoms of catecholamine excess. Phase II clinical trials with sunitinib, axitinib, and cabozantinib have revealed overall response rates of 13%, 36%, and 37%, respectively. Nevertheless, these medications could be associated with severe cardiovascular toxicity due to hypertension secondary to tumor lysis with subsequent release of catecholamines and direct TKI vascular toxicity [13, 14].

Cardiovascular toxicity has been substantial in the phase II trials with axitinib, lenvatinib, and pazopanib [15]. Patients with hormonally active mPPGL must be prepared with alpha- and beta-blockers; furthermore, the dose of the TKI must be individualized based on the patient's ability to tolerate side effects.

TKIs may be used for patients with rapid disease progression, and the toxicity of TKIs is expected to be lower than that of CVD chemotherapy. In addition, TKIs are a therapeutic option for patients with tumors which do not express the noradrenaline transporter, patients with mixed tumors (MIBG+/−), and patients with contraindications for radiopharmaceuticals, such as bone marrow insufficiency due to massive bone disease or prior treatments such as alkylating chemotherapy or radiopharmaceuticals. Although Food and Drug Administration (FDA) regulations in the USA allow prescription of TKIs in routine clinical practice, further exploration of these medications through clinical trials is needed.

Selective TKIs may offer less toxicity and impressive oncologic responses in patients with druggable mutations. Although PPGLs are frequently caused by monogenic mutations, most are not druggable, and only MEN2 pheochromocytomas have pathogenic tyrosine kinase mutations. Recently, RET inhibitors were approved for the treatment of medullary thyroid cancer associated with activating mutations of the RET proto-oncogene [16]. Multiple endocrine neoplasia type 2 metastatic pheochromocytomas are rare [17, 18]; however, patients with these tumors may benefit from treatment with RET inhibitors. In our Institution, for our MEN2 patients affected by mPPGL and highly pretreated we obtain pralsetinib on a named-patient basis, reaching a radiological partial response.

Clinical trials are extremely important to evaluate novel therapies targeting pathways involved in the development of mPPGL. Hypoxia-inducible factor 2 (HIF-2) is crucial since the three main pathways of PPGL growth converge on this element. Belzutifan, a HIF-2α inhibitor, was recently approved by the FDA for the treatment of sporadic kidney cancer and Von Hippel-Lindau–related tumors such as kidney cancer, pancreatic neuroendocrine tumors, and hemangioblastomas [19]. This medication has been associated with impressive clinical benefit rates and minimal toxicity [20]. Belzutifan is now being evaluated in a recently activated phase II clinical trial (MK6482 trial; ClinicalTrials.gov Identifier: NCT05239728) for patients with mPPGL. Clinical trials like this one may allow belzutifan to be considered as a first-line therapeutic option in the future.

17.3 Immunotherapy

Most mPPGLs are characterized by pseudohypoxia which contributes to the immune system's inability to recognize the disease, an important hallmark of cancer [21]. In a recent study, a substantial number of mPPGLs were found to express the programmed cell death ligands, making these cells potential targets of medications such as nivolumab and pembrolizumab [22]. A phase II clinical trial with pembrolizumab for patients with mPPGL revealed modest responses to pembrolizumab, with an overall response rate of 9% [23]. Patients tolerated treatment very well. The expression of PD-L1 and the presence of infiltrating mononuclear inflammatory cells in the primary tumor did not correlate with clinical responses. The authors of the study did not recommend single-agent pembrolizumab as first-line therapy for mPPGL; nevertheless, the results of the trial suggested potential mechanisms that could enhance the activity of immunotherapy. The simultaneous combination of immunotherapy with TKIs may lead to more impressive clinical responses since TKIs may release tumor antigens and induce an immunologic response; furthermore, TKIs may lead to vascular normalization that facilitates immune recognition [24].

A novel phase IB/II clinical trial (Spencer trial, EO2401; ClinicalTrials.gov Identifier: NCT04187404) combines nivolumab with a vaccine that contains several tumor antigens derived from the gut microbiome and presenting high affinity with adrenal tumor antigens. The vaccine may facilitate immune system recognition of mPPGL. Recently, at the last European Society for medical Oncology (ESMO) 2022 conference, preliminary data of 13 patients enrolled were presented showing a modest activity of nivolumab plus vaccine. Nine of 13 were previously treated, the overall response rate was 8%, disease control rate (overall response rate added to stable disease cases) 77% while mPFS reached 5.2 months and mOS 14.3 months.

References

1. Hescot S, Leboulleux S, Amar L, et al. One-year progression-free survival of therapy-naive patients with malignant pheochromocytoma and paraganglioma. J Clin Endocrinol Metab. 2013;98(10):4006–12.
2. Ayala-Ramirez M, Feng L, Habra MA, et al. Clinical benefits of systemic chemotherapy for patients with metastatic pheochromocytomas or sympathetic extra-adrenal paragangliomas: insights from the largest single-institutional experience. Cancer. 2012;118(11):2804–12.
3. Fishbein L, Del Rivero J, Else T, et al. The North American Neuroendocrine Tumor Society consensus guidelines for surveillance and management of metastatic and/or unresectable pheochromocytoma and paraganglioma. Pancreas. 2021;50(4):469–93.
4. Niemeijer ND, Alblas G, van Hulsteijn LT, et al. Chemotherapy with cyclophosphamide, vincristine and dacarbazine for malignant paraganglioma and pheochromocytoma: systematic review and meta-analysis. Clin Endocrinol (Oxf). 2014;81(5):642–51.
5. Keiser HR, Goldstein DS, Wade JL, et al. Treatment of malignant pheochromocytoma with combination chemotherapy. Hypertension. 1985;7(3 Pt 2):I18–24.
6. Tanabe A, Naruse M, Nomura K, et al. Combination chemotherapy with cyclophosphamide, vincristine, and dacarbazine in patients with malignant pheochromocytoma and paraganglioma. Horm Cancer. 2013;4(2):103–10.

7. Thosani S, Ayala-Ramirez M, Román-González A, et al. Constipation: an overlooked, unmanaged symptom of patients with pheochromocytoma and sympathetic paraganglioma. Eur J Endocrinol. 2015;173(3):377–87.
8. Guillemot J, Compagnon P, Cartier D, et al. Metoclopramide stimulates catecholamine- and granin-derived peptide secretion from pheochromocytoma cells through activation of serotonin type 4 (5-HT4) receptors. Endocr Relat Cancer. 2009;16(1):281–90.
9. Hadoux J, Favier J, Scoazec JY, et al. SDHB mutations are associated with response to temozolomide in patients with metastatic pheochromocytoma or paraganglioma. Int J Cancer. 2014;135(11):2711–20.
10. Tong A, Li M, Cui Y, Ma X, Wang H, Li Y. Temozolomide is a potential therapeutic tool for patients with metastatic pheochromocytoma/paraganglioma—case report and review of the literature. Front Endocrinol (Lausanne). 2020;11:61.
11. Fishbein L, Ben-Maimon S, Keefe S, et al. SDHB mutation carriers with malignant pheochromocytoma respond better to CVD. Endocr Relat Cancer. 2017;24(8):L51–5.
12. Huang L, Jiang S, Shi Y. Tyrosine kinase inhibitors for solid tumors in the past 20 years (2001–2020). J Hematol Oncol. 2020;13(1):143.
13. Jimenez C, Fazeli S, Román-Gonzalez A. Antiangiogenic therapies for pheochromocytoma and paraganglioma. Endocr Relat Cancer. 2020;27(7):R239–54.
14. O'Kane GM, Ezzat S, Joshua AM, et al. A phase 2 trial of sunitinib in patients with progressive paraganglioma or pheochromocytoma: the SNIPP trial. Br J Cancer. 2019;120(12):1113–9.
15. Jasim S, Suman VJ, Jimenez C, et al. Phase II trial of pazopanib in advanced/progressive malignant pheochromocytoma and paraganglioma. Endocrine. 2017;57(2):220–5.
16. Subbiah V, Hu MI, Wirth LJ, et al. Pralsetinib for patients with advanced or metastatic RET-altered thyroid cancer (ARROW): a multi-cohort, open-label, registrational, phase 1/2 study. Lancet Diabetes Endocrinol. 2021;9(8):491–501. Erratum in: Lancet Diabetes Endocrinol. 2021;9(10):e4.
17. Thosani S, Ayala-Ramirez M, Palmer L, et al. The characterization of pheochromocytoma and its impact on overall survival in multiple endocrine neoplasia type 2. J Clin Endocrinol Metab. 2013;98(11):E1813–9.
18. Castinetti F, Waguespack SG, Machens A, et al. Natural history, treatment, and long-term follow up of patients with multiple endocrine neoplasia type 2B: an international, multicentre, retrospective study. Lancet Diabetes Endocrinol. 2019;7(3):213–20. Erratum in: Lancet Diabetes Endocrinol. 2019;7(3):e3.
19. Choueiri TK, Bauer TM, Papadopoulos KP, et al. Inhibition of hypoxia-inducible factor-2α in renal cell carcinoma with belzutifan: a phase 1 trial and biomarker analysis. Nat Med. 2021;27(5):802–5. Erratum in: Nat Med. 2021;27(10):1849.
20. Hasanov E, Jonasch E. MK-6482 as a potential treatment for von Hippel-Lindau disease-associated clear cell renal cell carcinoma. Expert Opin Investig Drugs. 2021;30(5):495–504.
21. Jimenez C. Treatment for patients with malignant pheochromocytomas and paragangliomas: a perspective from the hallmarks of cancer. Front Endocrinol (Lausanne). 2018;9:277.
22. Pinato DJ, Black JR, Trousil S, et al. Programmed cell death ligands expression in phaeochromocytomas and paragangliomas: relationship with the hypoxic response, immune evasion and malignant behavior. Oncoimmunology. 2017;6(11):e1358332.
23. Jimenez C, Subbiah V, Stephen B, et al. Phase II clinical trial of pembrolizumab in patients with progressive metastatic pheochromocytomas and paragangliomas. Cancers (Basel). 2020;12(8):2307.
24. Economides MP, Shah AY, Jimenez C, et al. A durable response with the combination of nivolumab and cabozantinib in a patient with metastatic paraganglioma: a case report and review of the current literature. Front Endocrinol (Lausanne). 2020;11:594264.

Role of Radiotherapy in Adrenocortical Carcinoma and Pheochromocytoma

18

Marco Lorenzo Bonù and Stefano Maria Magrini

18.1 Role of Radiotherapy in Adrenocortical Carcinoma

18.1.1 Radiobiology of Adrenocortical Carcinoma

Classically, adrenocortical carcinoma (ACC) was considered a radioresistant disease. This assumption was based on results from very small series in which radiotherapy (RT) was used for palliation. More recent studies report improved tumor control being achieved by delivering higher doses using more accurate techniques. Such data suggest that ACC could be sensitive to higher doses and high doses per fraction. Thus, RT may have a role beyond palliation in adjuvant, advanced and recurrent disease. This chapter discusses the current role of RT in the management of ACC. A focus on the technique will help clinicians to familiarize with the lexicon of the radiation oncologist.

18.1.2 Adjuvant Radiotherapy

18.1.2.1 Aim of Radiotherapy

RT after curative-intent surgery uses high-energy ionizing radiation delivered to the tumor bed to prevent disease relapse by killing microscopically persistent cancer cells. The aim of adjuvant RT in ACC is therefore to reduce the risk of local relapse. Data on efficacy are mainly based on retrospective reports that show a benefit in local control for high-risk patients [1–3].

M. L. Bonù (✉) · S. M. Magrini
Radiotherapy, Department of Medical and Surgical Specialties, Radiological Sciences, and Public Health, University of Brescia at ASST Spedali Civili, Brescia, Italy
e-mail: marco.bonu@unibs.it; magrini@med.unibs.it

© The Author(s) 2025
G. A. M. Tiberio (ed.), *Primary Adrenal Malignancies*, Updates in Surgery,
https://doi.org/10.1007/978-3-031-62301-1_18

18.1.2.2 Patient Selection

RT is indicated for patients with high-risk features such as:

- non-curative R2 resection, at any stage of disease;
- margin positive R1/Rx resection, at any stage of disease;
- intraoperative violation of tumor capsule, tumor spillage or necrotic tumoral fluid dissemination, at any stage of disease;
- stage ≥III;
- R0 resection of tumors with adverse features such as: diameter >8 cm, lymphovascular invasion, Ki67 >10% and high-grade tumor with >20 mitotic figures per 50 HPF.

18.1.2.3 Efficacy and Timing

RT effectiveness in preventing local relapse is supported by small retrospective series with inhomogeneous results, reporting relapse rates ranging from 5% to 31.3%. RT is recommended to start no later than 12 weeks after surgery.

18.1.2.4 Acute and Late Toxicity

Expected adverse events depend on the laterality of the target, extension of the radiation volume to the para-aortic nodes, the RT technique and schedule. Concerning acute toxicity, G1–2 nausea, anorexia, abdominal pain and dyspepsia are frequent, with symptoms resolving a few days after the end of treatment owing to the close relationship with the stomach and duodenum. G1–2 fatigue is frequent and resolves after treatment. Late toxicity is rare, but G1–2 impaired kidney function with increased serum creatinine has been described. Radiation-induced liver disease and biliary tract disease have also been reported. Thanks to radiobiologic and technical advances, modern RT minimizes exposure of healthy tissues, making hepatic and biliary toxicities anecdotal [4].

18.1.2.5 Radiotherapy Technique, Dose and Volumes

The major body of evidence supporting the role of RT in the adjuvant scenario refers to 3D conformal RT (3DcRT) with X-ray photons. Briefly, 3DcRT delivers the dose to the target using multileaf collimators and multiportal fields which guarantee better dose conformity than older 2D techniques, potentially reducing the dose to healthy tissues close to the target. Nevertheless, contemporary RT uses intensity-modulated RT (IMRT) in its various declinations, such as: step-and-shoot IMRT, dynamic IMRT, helical IMRT, and volumetric-modulated arc therapy (V-MAT) (Fig. 18.1). Of note, stereotactic RT (SRT-SBRT) is a treatment modality were IMRT in its various declinations is used to precisely deliver a high radiation dose to a relatively small target, allowing high conformity to the target and steep dose fall-off outside, therefore maximizing efficacy to the target and minimizing the dose to critical organs at risk.

Modern techniques are important but not sufficient for performing high quality therapy on adrenal targets. In fact, respiratory motion management through tumor gating, tumor tracking, or breath holding is essential to maximize treatment

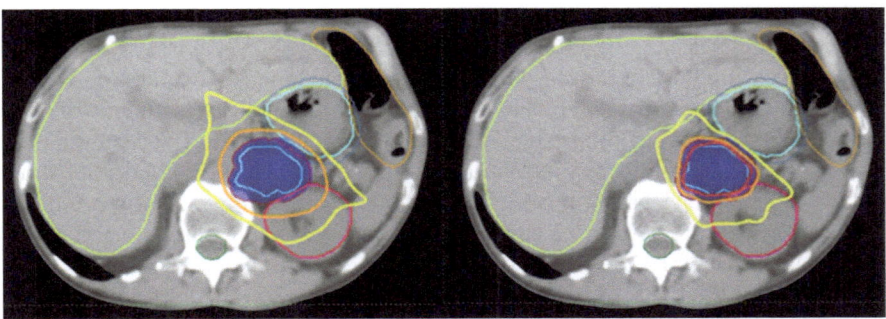

Fig. 18.1 Comparison between 3D conformal radiotherapy (3DcRT, *left*) and volumetric-modulated arc therapy (V-MAT, *right*) for the same dose prescription: 45 Gy in 6 fractions for an unresectable adrenocortical carcinoma. The solid blue area encompassed by the thin red line is the adrenal target volume, accounting for set-up errors (planning target volume, PTV); the thin sky blue line is the gross tumor volume, accounting for its motion during breathing (internal target volume, ITV); the yellow bold line represents 32 Gy isodose in 6 fractions, a dose potentially related to stomach damage; the orange bold line is the isodose prescription of 45 Gy in 6 fractions (a potentially therapeutic dose); the red bold line is 50 Gy isodose in 6 fractions. Note the higher conformity of the 32 and 45 Gy isodoses in the V-MAT plan (*orange* and *yellow bold lines*), the superior sparing of the stomach and the presence of 50 Gy isodose in the V-MAT plan that is absent in the 3DcRT plan. Therefore, careful V-MAT planning increases the doses to the target while minimizing doses to the stomach and other organs at risk

accuracy. These techniques have a critical role in further reducing the exposure of critical organs at risk; as a consequence, the toxicity profile of published series might be even better with the use of IMRT in its various declinations (including the SRT technique) without compromising the dose to the target.

18.1.2.6 Concurrent Systemic Therapy to Radiotherapy
There are limited data to guide decisions concerning the administration of mitotane concurrently to RT. Concurrent therapy is feasible without exceeding a mitotane dose of 3 g/day. An increase in abdominal RT-induced toxicity is expected when mitotane is administered during RT. Careful weekly monitoring of liver and kidney function is warranted, especially in right-sided irradiation.

18.1.3 Definitive Radiotherapy for Unresectable Disease or Local Recurrence

Use of RT to manage unresectable local disease is limited to very small single-center series. The literature up to 2022 reports on patients with an unresectable adrenal mass or local relapse in the tumor bed receiving RT with doses ranging between 39.2–73.5 Gy in 22–40-day schedules. The treatment techniques used 2D and 3DcRT. Despite these limitations, the responses achieved were encouraging and deserve some biological and technical considerations. From the biological point of view, ACC in the metastatic and palliative setting demonstrates a higher-than-expected radiosensitivity. From the technical point of view, contemporary techniques achieve a dose reduction to critical organs at risk while potentially increasing dose conformity to the target. Interestingly, several reports have described the feasibility, safety and effectiveness of high dose treatments of large adrenal masses using cutting-edge RT techniques, such as IMRT and proton therapy. A very large multi-institutional series published in 2023 confirmed those results: 132 target lesions in 80 patients were treated with modern techniques between 2010 and 2020. In 22 cases the target was a local recurrence and in 110 a metastasis. The RT schedules included normally fractionated RT (i.e., 2 Gy per fraction) with total doses ranging from 20 to 60 Gy and SRT. Considering the 22 patients treated for local recurrence, a complete response was observed in 13.6% of cases, a partial response in 36.4%, stable disease in 40.9%, while local progression occurred in 9.1% of cases. Median time to progression was 9.8 months. Such promising results form the backbone for a possible future role of RT in the multimodal management of unresectable/recurrent ACC [4].

18.1.4 Radiotherapy for Metastatic Disease and Palliation

RT is an option for the palliation of symptoms from locally advanced or distant metastatic disease, including metastatic bone pain.

The available literature supports the moderate benefit of palliative RT for symptom relief. In an old series of patients treated with moderate-low doses, a benefit in reduction of pain and other symptoms was achieved in 57% of patients, while more recent series describe better results with the use of higher doses, also in the palliative context [5, 6]. Such results support the hypothesis of the relative radiosensitivity of ACC to hypofractionated regimens and underline the importance of using contemporary techniques.

18.2 Role of Radiotherapy in Pheochromocytoma

18.2.1 Radiobiology of Pheochromocytoma

Pheochromocytoma (PHEO) was also considered a radioresistant disease. Despite the paucity and heterogeneity of published data, which also include paragangliomas in the analysis, it is now clear that a dose response to RT does exist and that doses >40 Gy (physical dose) seem associated with better local disease control in the context of metastatic disease [5].

Nevertheless, there are several other important issues that hampered the use of RT in this context. First, the early reports that identified a dose-response relationship also showed high toxicity to organs at risk. Such an observation is mainly explained by the attempt to reach dose escalation with outdated techniques (such as 2D techniques). Another important point is the unique biologic nature of PHEO. The production of catecholamines/metanephrines has raised concerns about the safety of RT, owing to the risk of a hypertensive crisis consequent to amine release after tumor cell death. Importantly, a possible exacerbation of hypertensive crisis few hours after the fifth day of palliative RT (20 Gy in 5 fractions) was described in only one case report. To date, no other hypertensive crisis triggered by tissue irradiation have been reported. On the other hand, no study has been designed to consistently test catecholamine/metanephrine changes before and after RT. Despite encouraging outcomes in modern series and the lack of G3 or worse side effects, this topic is still an open issue, and we suggest testing metanephrines and chromogranin-A before and shortly after RT.

Finally, as reported for other diseases, interpreting response to RT is challenging. PHEOs are characterized by slow response to RT and anatomical imaging may usually show residual masses also in the event of successfully treated lesions. An explanation for this behavior can be found in the heterogeneous tumor microenvironment, where tumor cells in active replication represent a minority of the tumor bulk. Magnetic resonance imaging, functional imaging together with biochemical response are useful in the differential diagnosis between disease persistence and tumor response [6].

18.2.2 Radiotherapy for Metastatic Disease and Palliation

In the past, the use of high-dose RT was associated with significant morbidity, which limited its utilization in malignant PHEO. More recently, reports of cases in which contemporary techniques (e.g., IMRT and SRT/SBRT) were employed described high dose delivery without significant toxicity to normal tissue resulting in a high rate of symptomatic and radiographic local disease control. However, most of these reports are biased by the limited number of patients and their selection. In fact, the available series analyzed together both PHEO and paraganglioma with different sites of metastatic spread, and delivered RT concurrently to radiometabolic therapy. As a consequence, there is wide heterogeneity in RT series for metastatic PHEO that limits the interpretation and applicability of results. Nevertheless, Vogel et al. treated 24 patients (13 with metastatic PHEO) on 36 metastatic sites (bone, pelvis, brain, upper abdomen) and found an overall symptomatic improvement in 81.1% of patients after RT regardless of site or radiation technique. Overall local control was equal to 86.7% in patients treated with mean doses of 31.8 Gy in 3.3-Gy fractions by 3DcRT and 21.9 Gy in 10.4-Gy fractions by SBRT/SRT. One case of G3 acute neuropathy emerged. No other acute or late ≥G3 toxicity was recorded [5]. Breen et al. in 2017 reported the outcomes of 41 patients (15 treated for PHEO on 37 lesions), confirming the fairly good results in terms of local control (81% at 5 years) and a dose-response relationship favoring patients treated with higher doses. Two patients developed grade ≥3 late adverse events thought to be related to RT (one case of iatrogenic menopause and one of sciatic neuropathy). Interestingly, seven patients were offered comprehensive treatment on all metastatic sites; a biochemical response was appreciated in all of the five patients who underwent blood/urine catecholamine/metanephrine assays before and after radiation [6].

Taken together, the limited available literature shows that RT is an effective treatment modality to control metastatic foci of PHEO, with most patients experiencing radiographic local tumor control and/or improvement of tumor-related symptoms. RT is also well tolerated, with few severe treatment-related adverse events and without hypertensive crises in modern series. Higher doses may guarantee improved local tumor control in selected patients.

18.2.3 Adjuvant Radiotherapy and Definitive Radiotherapy (Unresectable Disease)

There are no data concerning the use of RT as adjuvant treatment after resection or in the context of unresectable disease. Therefore, the authors of this chapter consider such an approach investigational and suggest that RT may be employed only within a clinical trial.

References

1. Fassnacht M, Dekkers OM, Else T, et al. European Society of Endocrinology Clinical Practice Guidelines on the management of adrenocortical carcinoma in adults, in collaboration with the European Network for the Study of Adrenal Tumors. Eur J Endocrinol. 2018;179(4):G1–G46.
2. Nelson DW, Chang SC, Bandera BC, et al. Adjuvant radiation is associated with improved survival for select patients with non-metastatic adrenocortical carcinoma. Ann Surg Oncol. 2018;25(7):2060–6.
3. Wu K, Liu X, Liu Z, et al. Benefit of postoperative radiotherapy for patients with nonmetastatic adrenocortical carcinoma: a population-based analysis. J Natl Compr Cancer Netw. 2021;19(12):1425–32.
4. Kimpel O, Schindler P, Schmidt-Pennington L, et al. Efficacy and safety of radiation therapy in advanced adrenocortical carcinoma. Br J Cancer. 2023;128(4):586–93.
5. Vogel J, Atanacio AS, Prodanov T, et al. External beam radiation therapy in treatment of malignant pheochromocytoma and paraganglioma. Front Oncol. 2014;4:166.
6. Breen W, Bancos I, Young WF Jr, et al. External beam radiation therapy for advanced/unresectable malignant paraganglioma and pheochromocytoma. Adv Radiat Oncol. 2017;3(1):25–9.

Radionuclide Treatment in Malignant Pheochromocytoma

19

Francesco Dondi and Francesco Bertagna

19.1 Introduction

Metastatic pheochromocytomas and paragangliomas (mPPGLs) can have heterogeneous behavior: some are very aggressive with rapid growth while others are asymptomatic with minimal or no progression. Nevertheless, most mPPGLs will at some point require systemic therapy [1–4].

Theragnostics is a field of nuclear medicine that focuses on the therapeutic and diagnostic capabilities of a single pharmacological platform: the likelihood of benefit from targeted radionuclide therapies could be accurately determined by imaging findings using the same radiopharmaceutical. $^{123/131}$I-metaiodobenzylguanidine (MIBG) and somatostatin receptor (SSTR) targeted with ^{90}Y-, ^{68}Ga-, and ^{177}Lu-somatostatin analogs are used for mPPGL radio-theragnostics [2, 4–6].

19.2 ^{131}I-MIBG Therapy

MIBG is an analog of guanethidine which has neuroendocrine transporter (NET) uptake and accumulates in neurosecretory granules of the sympathetic presynaptic neurons of mPPGLs [1, 3, 7]. In this setting, ^{123}I-MIBG is used to assess NET expression by mPPGLs in order to enable targeted radionuclide therapy, since it has superior imaging characteristics compared to therapeutic ^{131}I-MIBG [5, 6, 8]. In fact, ^{123}I-MIBG uptake is seen in 92% of pheochromocytomas and 64% of paragangliomas [3, 7] (Fig. 19.1). In general, ^{131}I-MIBG achieved complete response (CR) in 3% of patients, partial response (PR) in 27% subjects and stable disease (SD) in 52% [6, 9, 10] (Fig. 19.2).

F. Dondi (✉) · F. Bertagna
Nuclear Medicine, Department of Medical and Surgical Specialties, Radiological Sciences, and Public Health, University of Brescia at ASST Spedali Civili di Brescia, Brescia, Italy
e-mail: f.dondi@outlook.com; francesco.bertagna@unibs.it

© The Author(s) 2025
G. A. M. Tiberio (ed.), *Primary Adrenal Malignancies*, Updates in Surgery,
https://doi.org/10.1007/978-3-031-62301-1_19

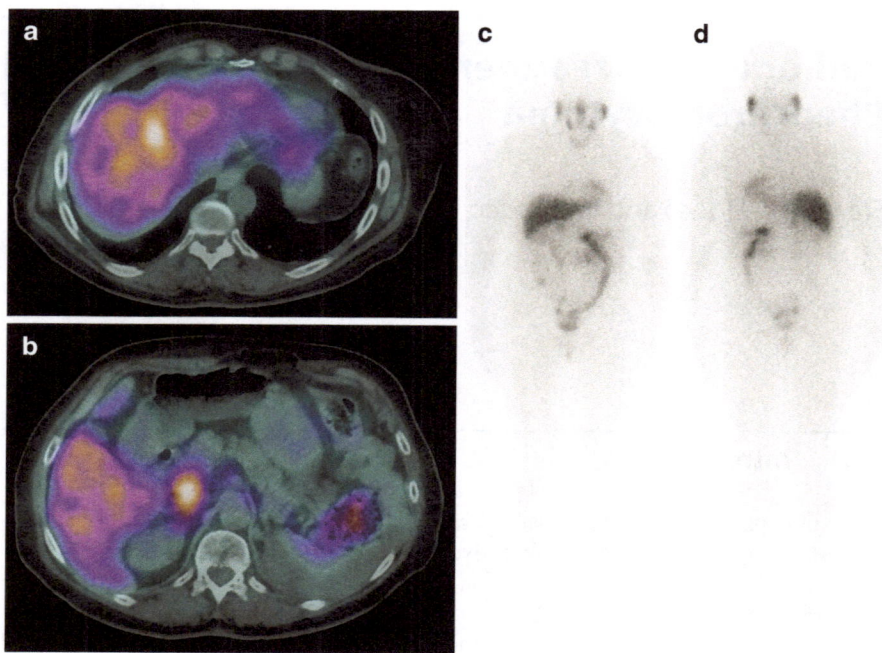

Fig. 19.1 ^{123}I-MIBG SPECT/CT (**a**, **b**) and planar scintigraphic (**c**, **d**) images of a patient affected by pheochromocytoma with hepatic and portacaval nodal metastatic lesions

Two types of ^{131}I-MIBG are available: a low-specific-activity (LSA) formulation and a high-specific-activity (HSA) formulation. In the first, more than 99% of MIBG molecules are unlabeled (15–50 mCi/mg), whereas the HSA formulation contains a larger amount of labeled molecules (~2500 mCi/mg), reducing side effects and competitiveness with the unlabeled MIBG [2, 11].

19.2.1   HSA ^{131}I-MIBG

The HSA ^{131}I-MIBG formulation is approved for the treatment of patients aged 12 years and older with progressive and unresectable MIBG-avid mPPGL. Moreover, it should be considered as a first-line approach when systemic therapy is required to achieve disease stabilization or symptom control [1, 3, 12, 13].

The HSA regimen incorporates an initial dosimetric study using ^{131}I-MIBG. For therapeutic purposes, a total of two doses of 500 mCi (or 8 mCi/kg if weight is <62.5 kg) are infused intravenously at least 90 days apart; the infusion is administered over 30 min in adults and 60 min in children. All patients need pretherapy thyroid blockade with potassium iodide to avoid hypothyroidism (130 mg 24–48 h before therapy, continued for 10–15 days) [1, 6–8]. Medications which can affect catecholamine uptake should be stopped for at least 5 half-lives before and 7 days

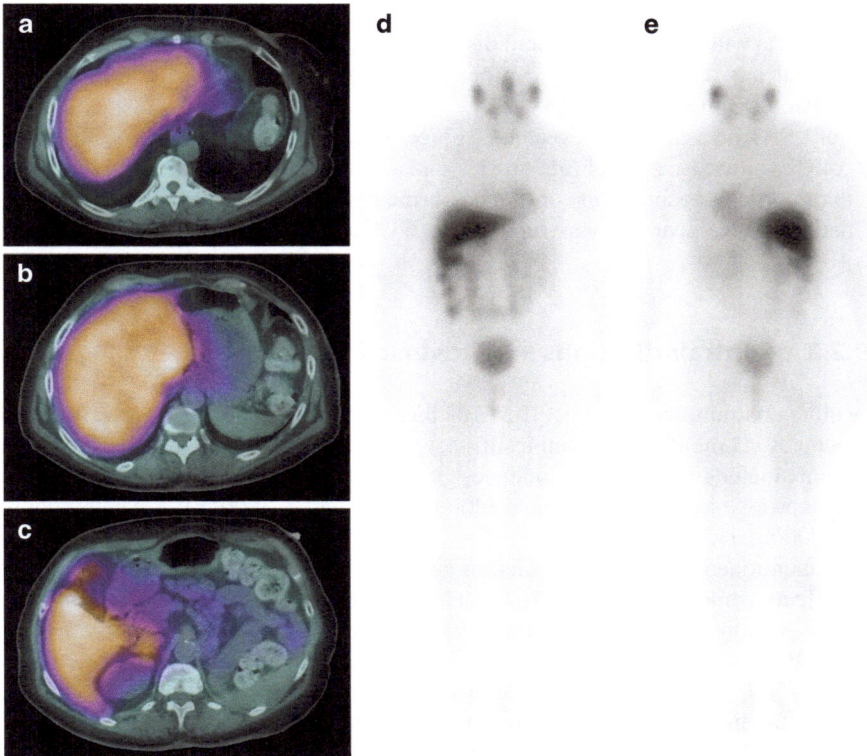

Fig. 19.2 ^{131}I-MIBG SPECT/CT (**a–c**) and planar scintigraphic (**d, e**) images of the same patient of Fig. 19.1 performed after targeted radionuclide therapy, confirming the presence of hepatic metastases but only faint nodal uptake

after therapy. In patients with mild to moderate renal impairment a dose reduction may be required [7].

One trial reported a reduction in baseline antihypertensive medication in 25% of patients with mPPGL, PR in 23% and SD in 69% of the subjects. At 1-year follow-up, 68% of them had PR confirmation or CR [11]. The median overall survival (OS) was 17.5 months after a single therapeutic dose and 48.7 months after two doses. Other studies reported somewhat higher objective response rates as well as a reduction of serum tumor marker levels [4, 6–8, 14].

19.2.2   LSA ^{131}I-MIBG

No clear and approved regimens exist for LSA ^{131}I-MIBG, and the doses used range from 50 to 3200 mCi over 1–12 administrations. Since LSA ^{131}I-MIBG contains a high mass of unlabeled radiopharmaceuticals, in the case of hypertension developing, it may be necessary to pause or decrease the infusion rate [7–9, 15]. Thyroid blockade and discontinuation of medications influencing MIBG uptake are required. For subjects with relatively indolent disease or unwilling to undergo in-patient therapy, serial low dose treatments can be considered (2–3 mCi/kg or 200 mCi/cycle administered 3 months apart) [8].

For LSA, meta-analyses revealed an objective radiological response of 30% with 4% of CR, a disease control rate of 82% and a biochemical response of 51%; complete or partial catecholamine or metanephrine response was observed in 19–100% of patients. Five-year OS was reported as 64% and event-free survival was 47% [8, 10].

19.2.3 Contraindications and Adverse Effects

Absolute contraindications for ^{131}I-MIBG therapy are pregnancy, breastfeeding, life expectancy <3 months and renal insufficiency requiring dialysis. Relative contraindications include urinary incontinence, glomerular filtration rate <30 mL/min and bone marrow suppression (white blood cell count <3000/mL, platelet count <100,000/mL) [5, 8].

As mentioned, the LSA formulation can have a higher rate of pharmacological side effects compared to HAS [7, 11]. Hematological toxicity (thrombocytopenia, anemia, leukopenia and neutropenia) is considered the most severe side effect of ^{131}I-MIBG therapy, but it is either self-limiting or treatable with therapeutic intervention [5, 7, 8, 11]. Non-hematological toxicities can include nausea, vomiting, fatigue, and anorexia (4–49% of cases), usually beginning a few days after administration, persisting for 3–4 weeks and being self-limiting or pharmaceutically treatable. In the case of the HSA regimen, there may be a decline in renal function with renal failure or acute kidney injury [7]. Hypertensive crises after therapy infusion can occur, in particular for high doses of the LSA formulation. Catecholamine blockade with α- and β-blockers is suggested in these cases. Other catecholamine release symptoms are possible during MIBG infusion or in the early post-treatment period [5, 7, 8, 11]. Secondary malignancies (acute myeloid leukemia, chronic myeloid leukemia, and myelodysplastic syndrome) have been documented years after ^{131}I-MIBG therapy. Hypothyroidism may be observed in the absence of thyroid blockade [5, 7, 8].

19.3 Peptide Receptor Radionuclide Therapy

^{68}Ga-labeled DOTA peptides (DOTANOC, DOTATOC, and DOTATATE) allow the evaluation of SSTR expression of mPPGLs with PET/CT, enabling the selection of patients who can benefit from peptide receptor radionuclide therapy (PRRT) [2, 5, 6]. In this setting, when the ligand is changed to ^{177}Lu or ^{90}Y the tracers gain the ability to emit β-radiation and therefore act as a therapeutic agent.

19.3.1 ^{90}Y- and ^{177}Lu-Labeled DOTA Compounds

For PRRT applied to mPPGL various radiotracers are available: ^{177}Lu-DOTATATE (Lu-PRRT), ^{90}Y-DOTATOC (Y-PRRT), and ^{90}Y-DOTATATE, even if the most used is the first one. Some insights suggested that patients who received Lu-PRRT had a longer OS than those receiving Y-PRRT, but many studies also showed the therapeutic benefit of Y-PRRT [1, 16]. Different protocols are available for Y-PRRT (typically 30 mCi in 5 cycles with a 1–11 range) and for Lu-PRRT (typically 200 mCi in 5 cycles) [1, 11, 17].

In general, different meta-analyses reported that 90% of the patients achieved PR or SD, with an objective response rate of 25%, a disease control rate (DCR) of 84%, a clinical response of 61% and a biochemical response of 64% [8, 11, 18, 19]. Interestingly, studies with Lu-PRRT reported an overall response rate (ORR) of 26% and a DCR of 83%, while Y-PRRT was found to have pooled ORR and DCR of 24% and 85%, respectively [6]. Moreover, for ^{90}Y-DOTATATE, studies reported a PR of 8%, a SD of 75%, and a progressive disease (PD) of 17% at 6 months, and no PR, a SD of 82%, and a PD of 18% at 12 months [6].

19.3.2 Contraindications and Adverse Effects

Absolute contraindications for PRRT include pregnancy and the presence of serious concurrent diseases or unmanageable psychiatric disorders. Relative contraindications include breastfeeding, impaired renal function, red blood cell count <3,000,000/mL, white blood cell count <3000/mL, absolute neutrophil count <1000/mL, and platelet count <75,000/mL [5]. Adequate liver function should be documented [8].

Patients with mPPGL are at high risk of catecholamine release syndrome, in particular when receiving Lu-PRRT [1, 8, 11]. Common acute events include nausea and vomiting, which can be relieved with a continuous infusion L-lysine or L-arginine [8].

The kidneys and bone marrow are the dose-limiting organs for PRRT, although severe toxicity reactions have rarely been observed: neutropenia in 3% of cases, thrombocytopenia in 9%, lymphopenia in 11% and nephrotoxicity in 4–9% [5, 11, 16, 18]. The general incidence of myelodysplasia has been reported in 2–8% [8]. Renal protection is provided by concomitant amino acid infusion to block reabsorption of the radiopharmaceuticals in the renal tubes [11].

19.4 Comparison Between ^{131}I-MIBG Therapy and Peptide Receptor Radionuclide Therapy

Different studies compared ^{131}I-MIBG and PRRT in the treatment of mPPGL and the choice between these two therapies is driven by the relative uptake of tracers on imaging scans. Interestingly, mPPGLs generally have higher expression of SSTR than NET, as suggested by the higher sensitivity of PET imaging [7]. With either agent, an important step is to consider the toxicity profile and the patient's characteristics, in particular bone marrow reserve and potential development of acute catecholamine release syndrome [1, 11]. In this setting, ^{131}I-MIBG has a lower risk of catecholamine crises and should be considered in patients with good bone marrow reserve [5, 11].

In a comparison between Lu-PRRT and ^{131}I-MIBG, the biochemical response (100% vs. 50% respectively), objective response (44% vs. 17%), DCR (100% vs. 83%), symptom control (87% vs. 75%) and progression-free survival (PFS) (29 vs. 19–25 months) were worse for the second regimen [5, 6]. Interestingly, ^{131}I-MIBG showed longer PFS than Lu-PRRT when the proportion of pheochromocytomas in the cohort was low [2].

References

1. Jimenez C, Xu G, Varghese J, et al. New directions in treatment of metastatic or advanced pheochromocytomas and sympathetic paragangliomas: an American, contemporary, pragmatic approach. Curr Oncol Rep. 2022;24(1):89–98.
2. Prado-Wohlwend S, Del Olmo-García MI, Bello-Arques P, Merino-Torres JF. Response to targeted radionuclide therapy with [131I]MIBG AND [177Lu]Lu-DOTA-TATE according to adrenal vs. extra-adrenal primary location in metastatic paragangliomas and pheochromocytomas: a systematic review. Front Endocrinol (Lausanne). 2022;13:957172.
3. Wang K, Crona J, Beuschlein F, et al. Targeted therapies in pheochromocytoma and paraganglioma. J Clin Endocrinol Metab. 2022;107(11):2963–72.
4. Garcia-Carbonero R, Matute Teresa F, Mercader-Cidoncha E, et al. Multidisciplinary practice guidelines for the diagnosis, genetic counseling and treatment of pheochromocytomas and paragangliomas. Clin Transl Oncol. 2021;23(10):1995–2019.
5. Zhang X, Wakabayashi H, Hiromasa T, et al. Recent advances in radiopharmaceutical theranostics of pheochromocytoma and paraganglioma. Semin Nucl Med. 2023;53(4):503–16.
6. Jungels C, Karfis I. 131I-metaiodobenzylguanidine and peptide receptor radionuclide therapy in pheochromocytoma and paraganglioma. Curr Opin Oncol. 2021;33(1):33–9.
7. Dillon JS, Bushnell D, Laux DE. High-specific-activity 131iodine-metaiodobenzylguanidine for therapy of unresectable pheochromocytoma and paraganglioma. Future Oncol. 2021;17(10):1131–41.
8. Carrasquillo JA, Chen CC, Jha A, et al. Systemic radiopharmaceutical therapy of pheochromocytoma and paraganglioma. J Nucl Med. 2021;62(9):1192–9.
9. Gonias S, Goldsby R, Matthay KK, et al. Phase II study of high-dose [131I]metaiodobenzylguanidine therapy for patients with metastatic pheochromocytoma and paraganglioma. J Clin Oncol. 2009;27(25):4162–8.
10. van Hulsteijn LT, Niemeijer ND, Dekkers OM, Corssmit EP. (131)I-MIBG therapy for malignant paraganglioma and phaeochromocytoma: systematic review and meta-analysis. Clin Endocrinol (Oxf). 2014;80(4):487–501.
11. Jha A, Taïeb D, Carrasquillo JA, et al. High-specific-activity-131I-MIBG versus 177Lu-DOTATATE targeted radionuclide therapy for metastatic pheochromocytoma and paraganglioma. Clin Cancer Res. 2021;27(11):2989–95.
12. Fishbein L, Del Rivero J, Else T, et al. The North American Neuroendocrine Tumor Society consensus guidelines for surveillance and management of metastatic and/or unresectable pheochromocytoma and paraganglioma. Pancreas. 2021;50(4):469–93.

13. Fassnacht M, Assie G, Baudin E, et al. Adrenocortical carcinomas and malignant phaeochromocytomas: ESMO-EURACAN Clinical Practice Guidelines for diagnosis, treatment and follow-up. Ann Oncol. 2020;31(11):1476–90. Erratum in: Ann Oncol. 2023;34(7):631.
14. Pryma DA, Chin BB, Noto RB, et al. Efficacy and safety of high-specific-activity 131I-MIBG therapy in patients with advanced pheochromocytoma or paraganglioma. J Nucl Med. 2019;60(5):623–30.
15. Loh KC, Fitzgerald PA, Matthay KK, et al. The treatment of malignant pheochromocytoma with iodine-131 metaiodobenzylguanidine (131I-MIBG): a comprehensive review of 116 reported patients. J Endocrinol Investig. 1997;20(11):648–58.
16. Patel M, Tena I, Jha A, et al. Somatostatin receptors and analogs in pheochromocytoma and paraganglioma: old players in a new precision medicine world. Front Endocrinol (Lausanne). 2021;12:625312.
17. Severi S, Bongiovanni A, Ferrara M, et al. Peptide receptor radionuclide therapy in patients with metastatic progressive pheochromocytoma and paraganglioma: long-term toxicity, efficacy and prognostic biomarker data of phase II clinical trials. ESMO Open. 2021;6(4):100171.
18. Taïeb D, Jha A, Treglia G, Pacak K. Molecular imaging and radionuclide therapy of pheochromocytoma and paraganglioma in the era of genomic characterization of disease subgroups. Endocr Relat Cancer. 2019;26(11):R627–R52.
19. Satapathy S, Mittal BR, Bhansali A. Peptide receptor radionuclide therapy in the management of advanced pheochromocytoma and paraganglioma: a systematic review and meta-analysis. Clin Endocrinol (Oxf). 2019;91(6):718–27.

Preclinic and Translational Research in Adrenal Malignancies

20

Elena Rapizzi, Andrea Abate, Mariangela Tamburello, Michaela Luconi, and Sandra Sigala

20.1 Introduction

Preclinical studies in the field of adrenal cancer, both *in vitro* and *in vivo* using adequate and useful cell and animal models, are mandatory to study the molecular and cellular characteristics of these tumors as well as to investigate the potential of a therapeutic drug or strategy. Because results obtained at preclinical level are important steps before translation to clinical trials, such experiments must be designed, conducted, analyzed and reported to the highest levels of rigor and transparency. The principal experimental models of adrenocortical carcinoma (ACC), pheochromocytoma (PHEO) and paraganglioma (PGL) are briefly described, taking into consideration that this field is rapidly evolving, with preclinical models being developed and validated to reproduce the tumor microenvironment (TME).

E. Rapizzi
Department of Experimental and Clinical Medicine, University of Florence, Florence, Italy
Centro di Ricerca e Innovazione sulle Patologie Surrenaliche, Careggi University Hospital, Florence, Italy
e-mail: elena.rapizzi@unifi.it

A. Abate · M. Tamburello · S. Sigala (✉)
Department of Molecular and Translational Medicine, University of Brescia, Brescia, Italy
e-mail: andrea.abate@unibs.it; mariangela.tamburello@unibs.it; sandra.sigala@unibs.it

M. Luconi
Department of Experimental and Clinical Biomedical Sciences, University of Florence, Florence, Italy
Centro di Ricerca e Innovazione sulle Patologie Surrenaliche, Careggi University Hospital, Florence, Italy
e-mail: michaela.luconi@unifi.it

© The Author(s) 2025
G. A. M. Tiberio (ed.), *Primary Adrenal Malignancies*, Updates in Surgery,
https://doi.org/10.1007/978-3-031-62301-1_20

20.2 Adrenocortical Carcinoma: Preclinical and Translational Models

The heterogeneity of ACC cell phenotypes requires cellular models capable of reproducing this condition. Preclinical mouse models of ACC have been generated in the form of cell line-derived xenografts and patient-derived xenografts, established using cell suspension cultures or tumor tissues from surgery, respectively. These are administered subcutaneously to immunocompromised mice for localized propagation and *in vivo* tumor growth, offering the opportunity to observe *in vivo* the progression of the human tumor, however with some limitations [1]. Transgenic mouse models are also available but their genetic modifications might only partially recapitulate the heterogeneity of human disease [2] (Fig. 20.1).

20.2.1 Cell Lines

Currently, six human ACC-derived cell lines are used as models for this disease, and several primary cultures of ACC are used in preclinical studies. The principal characteristics of the human ACC cell models are described in depth in two recent reviews [2, 3]. The first human ACC cell line, named NCI-H295 derived from a

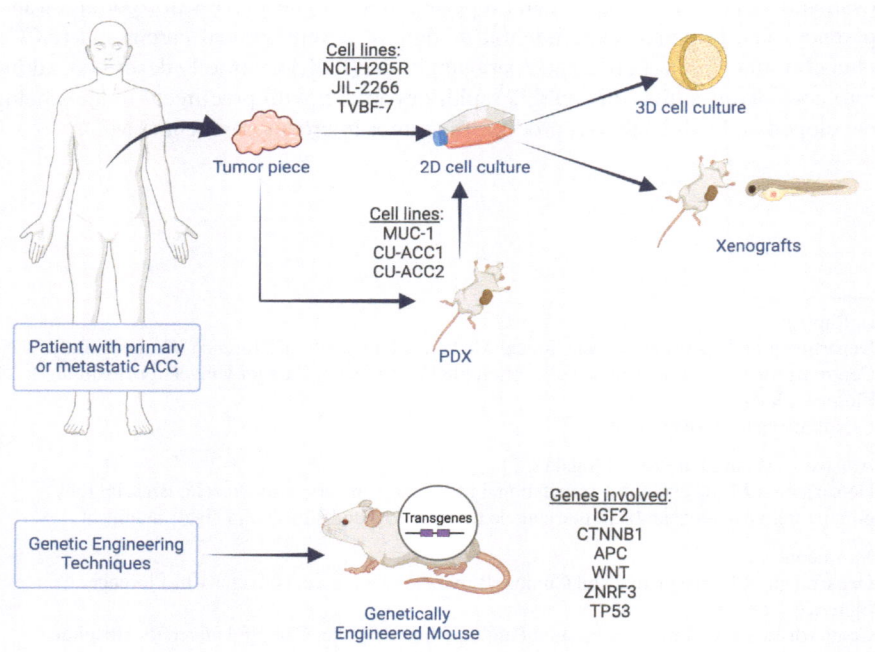

Fig. 20.1 Schematic representation of experimental models of adrenocortical carcinoma. (Created with BioRender.com)

primary ACC, was reported in 1990. Several sub-strains were adapted from the NCI-H295 cell line using alternative growth conditions, the first of which was NCI-H295R cells. NCI-H295R cells have been reported to harbor a large deletion in the *TP53* locus and carry an activating *CTNNB1* mutation. Furthermore, NCI-H295 cells and their sub-strains have been shown to produce steroids under basal conditions. Interestingly, the reported steroidogenic capacities are influenced by the culture conditions and the substrate [4].

The first metastasis-derived ACC cell line MUC-1 was established in 2016. MUC-1 cells represent a preclinical model of resistance to treatment with etoposide, doxorubicin, cisplatin and mitotane (EDP-M), as they were obtained from a patient progressing after EDP-M. MUC-1 cells are characterized by a low steroidogenic activity and by a somatic deletion/frameshift mutation in the *TP53* gene. Two other metastasis-derived ACC cell lines were subsequently established in 2018, namely CU-ACC1 and CU-ACC2. The CU-ACC1 cell line secretes high levels of cortisol but not aldosterone, and it is endowed with an activating point mutation of the *CTNNB1* gene. CU-ACC2 cells secrete very low amounts of cortisol and carry a mutation in *TP53*. Another recently available ACC cell model is the JIL-2266 cell line, derived from a primary ACC and first reported in 2021. JIL-2266 cells are characterized by intermediate to low expression of SF-1 and their hormone production depends on the composition of the culture medium. Genetically, JIL-2266 cells carry a hemizygous stop-gain mutation in the *TP53* gene. Furthermore, a pathogenic germline mutation in the *MUTYH* gene was observed, leading to the inactivation of the base excision-repair process and, consequently, to a high tumor mutational burden. The most recent ACC cell line, named TVBF-7, was established from lymph node metastasis of a patient progressing after EDP-M treatment. TVBF-7 cells produce high levels of cortisol under basal conditions. Genetic analysis reported an altered Wnt/β-catenin pathway due to the presence of a nonsense *APC* mutation.

Murine cell lines are also available. Among these, ATC1 and ATC7 cells were generated from transgenic mice and are used as cellular models in basic endocrinological studies [5, 6].

A different approach involves the use of co-cultures. The co-culture of a NCI-H295R cell monolayer above an adipose stem cell monolayer led to reprogramming of both cell types and to a more aggressive disease phenotype [7]. Recent experiments have also looked at interactions between ATC7 cells and human monocytes, showing that activation of intra-adrenal immune cells may play a role in stimulating steroidogenesis or proliferation [2].

20.2.2   3D Cell Models

3D models are particularly promising as an opportunity to better recapitulate the metabolic interplay between TME, cancer cells and tissue zonation. Indeed, 3D models replicate *in vivo* ACC tumor growth, since they have several important characteristics of solid tumors, including more representative transcriptional profiles, the development of an extracellular matrix and cellular junctions. Similar to *in vivo* solid tumors, there are various concentrations of oxygen and nutrients as well as different rates of cell proliferation from the outer layer to the center, which can result in central necrosis and regions of hypoxia. 3D models of ACC consisted mostly of spheroids generated from NCI-H295R cells primarily used in drug-screening protocols [8, 9], and they have been standardized in both NCI-H295R and MUC-1 cells as well as in ACC primary cultures [10].

Another research group released promising organoid models of ACC, studying metastasis through matrix metalloproteinase experiments in organoids and micro-fluidic models and demonstrating that NCI-H295R cells secrete active matrix metal-loproteinases [2].

20.2.3   Cell Xenograft in Mouse

Patient-derived xenografts (PDXs) in immunodeficient mice are recognized as the gold standard for human cancer models. However, limited PDXs as well as cell line-derived xenograft (CDX) models of ACC are available. The first PDX model reported in 2013 was generated from a pediatric patient with ACC. No separate cell line of this model has been established [2]. Since then, three new models have been developed, from which derive MUC-1, CU-ACC1 and CU-ACC2 cell lines, which not only retain significant molecular similarity to their primaries, but also recapitulate the differences between those primaries and some of the heterogeneity of the disease [11, 12]. Further work has investigated the behavior of one of these models, CU-ACC2-M2B, in a humanized mouse model to better understand the efficacy of checkpoint inhibitor immunotherapy. A detailed description of these models has been recently published [2].

In vivo experiments with the zebrafish offer, with some limitations, a suitable and expeditious animal model for the screening of potentially effective drugs, identification of dose toxicity, and determination of the most promising compounds for more advanced preclinical phases, especially in rare diseases such as ACC. Different studies reported results obtained with zebrafish embryos xenografted with ACC cells, to investigate the effect of different drugs on tumor growth and metastasis formation [13–15].

20.2.4   Genetically Engineered Mouse Models

Efficient mouse genome manipulation has allowed the development of genetically modified ACC models, engineered to contain specific genetic alterations which promote *de novo* tumor formation within the adrenal cortex; genes and pathways of interest are recognized by human clinical observations or *in vitro* findings. Once identified, a gene can be deleted, overexpressed, or mutated within the adrenal cortex to experimentally define its role in the pathogenesis of adrenal tumors [16]. Early models mostly focused on the role of IGF2 to elucidate its role in adrenocortical neoplasia. Other recent work has focused more on *CTNNB1*, *APC*, *WNT*, *ZNRF3*, and *TP53* [2, 17]. Val et al. recently showed that inactivation of *ZNRF3* in the mouse adrenal cortex, recapitulating the most frequent alteration in ACC patients, is associated with sexually dimorphic tumor progression, promoting or hampering the involvement of phagocytic macrophages in men and women, respectively [18].

20.3 Pheochromocytoma/Paraganglioma Preclinical and Translational Models

The need for cellular models of pheochromocytomas and paragangliomas (together referred to as PPGLs) is becoming urgent as basic research shifts from genetics to the molecular mechanisms driving the tumorigenesis and clinical vulnerabilities of PPGLs [19–21]. Research efforts have largely focused on *SDHB*, since the risk of metastasis is strongly related to genotype, ranging between 30% and 40% with *SDHB* mutations.

20.3.1 Cell Lines

PPGL cell line models include PC12, MPC cells, the MPC derivative MTT, the immortalized chromaffin cells, the putative human PHEO progenitor line hPheo1, and the recently developed RS0 and RS1/2 cell lines (Table 20.1).

Table 20.1 Characteristics of the different cell lines used in preclinical models of adrenocortical carcinoma and pheochromocytoma/paraganglioma

Cell line	Origin	Genotype Cluster	Catecholamine secretion	Forming clusters/ spheroids
PC12	Rat	*MAX* deletion Cluster 2	Norepinephrine	–
MPC	Mouse	Heterozygous *NF1* knockout Cluster 2	Norepinephrine, epinephrine	+
MTT	Metastatic mouse pheochromocytoma	Heterozygous *NF1* knockout Cluster 2	Norepinephrine, epinephrine	+
hPheo1	Human	*hTERT* immortalized Cluster 2	Not secreting	–
ImCC	Mouse	*SDHB* knockout Cluster 1	Unknown	–
RS1/2	Rat	Heterozygous *SDHB* knockout Cluster 1	As xenograft: dopamine, norepinephrine, epinephrine	Unknown
R50	Rat	*SDHB* knockout Cluster 1	As xenograft: dopamine, norepinephrine	+

20.3.2 Rat Pheochromocytoma (PC12)

The adrenal rat pheochromocytoma (PC12) cell line was originally isolated from a PHEO developed in an irradiated rat in 1976 [22], and subsequently found harboring *MAX* gene deletion [23]. This cell line has the characteristic of precursor cells for both sympathetic neurons and chromaffin cells. PC12 can differentiate toward a neuronal phenotype in response to the nerve growth factor, while dexamethasone treatment upregulates catecholamine synthesis and storage. PC12 expresses several of the catecholamine biosynthetic enzymes, including tyrosine hydroxylase along with the enzymes converting L-DOPA to dopamine (aromatic L-amino acid decarboxylase) and dopamine to norepinephrine (DA β-hydroxylase), while phenylethanolamine *N*-methyltransferase (PNMT), which produces epinephrine from norepinephrine, is not expressed in these cells. PC12 cells are now available from a wide range of cell repositories and have been used in many studies, not only related to adrenal function and catecholamine production, but also in neuronal differentiation and other aspects of neurological development and function.

20.3.3 Mouse Pheochromocytoma Cell Line and Mouse Tumor Tissue Cells

Mouse pheochromocytoma cell (MPC) and the later derived mouse tumor tissue (MTT) cells, developed in the labs of Arthur Tischler and Karel Pacak, respectively, were derived from PHEOs arising in the adrenal medulla of the *NF1* knockout mouse [24, 25]. The MPC cell line generally employed for preclinical studies pertaining to human tumors is 4/30/PRR. These cells typically showed extensive spontaneous neuronal differentiation. MPC and MTT are valid tools for studying genes and signaling pathways governing cell growth and differentiation in adrenal medullary neoplasms and are a unique model for studying the regulation of PNMT expression, as they display positive staining for PNMT and produce epinephrine. These cells are also considered a useful model for studying neurotransmitter release and neuroendocrine secretion. Several cell lines were then derived from MPC and MTT to improve and diversify research studies. To assess possible associations between *SDHB* gene mutations and invasiveness, Richter et al. established an MTT *SDHB* knockdown by viral transduction with lentiviral particles. Since MPC and MTT spontaneously form clusters in cultures, it is possible to easily generate spheroids [26], extremely useful for switching from monolayer (2D) cultures to 3D spheroids. Spheroids provide an excellent *in vitro* model to study the influence of hypoxia under conditions close to the *in vivo* situation, and for anticancer drug screening.

20.3.4 Immortalized Chromaffin Cells

Another mouse cell model, dubbed "immortalized mouse chromaffin cells" (imCCs), was derived from an *SDHB* knockout mouse [27]. These cells are deficient for the SDHB protein and show loss SDH activity, accompanied by high levels of intracellular and secreted succinate. Letouzé et al. also found other established characteristics of *SDHB* loss in imCCs, including elevated expression and nuclear translocation of *HIF2a* and a hypermethylation phenotype. Nevertheless, imCCs exhibit a mesenchymal morphology suggesting that they may not be mature chromaffin cells.

20.3.5 Rat *SDH*-Deficient RS0 Cells

PHEOs from irradiated rats with a heterozygous germline *SDHB* mutation were injected subcutaneously into NOD scid gamma (NSG) mice. This approach led to obtain two distinct, serially transplantable, xenograft and cell lines designated RS0 (for rat Sdh null) and RS1/2 (for rat Sdh haplo-insufficient). The ultrastructural features of RS0 are reminiscent of human SDH-deficient tumors, with relatively sparse secretory granules and cytoplasmic vacuoles, but the typical mitochondrial swelling and degeneration found in many human tumors are absent. The catecholamine profile of RS0 is also evocative of some *SDH*-deficient human PGLs, predominantly producing dopamine, with low levels of norepinephrine and undetectable epinephrine [28].

20.3.6 Progenitor Cells Derived from a Human Pheochromocytoma

In 2013, human primary PHEO cells were immortalized by introducing the catalytic subunit of human telomerase reverse transcriptase (hTERT) into the cells [29]. The resulting cell line is a neuroendocrine progenitor cell line called hPheo1. The characterization of these hPheo1 cells showed that the genes associated with catecholamine synthesis were highly expressed in the tumor tissue of origin, but most of them were downregulated in hPheo1 cells. More recently, genomic deletion of SDHB in hPheo1 cells was performed by the CRISPR/AsCPF1 system [30], obtaining a new cell line called hPheo1 *SDHB*-knockout.

20.3.7 Animal Models

In addition to the rat model harboring a heterozygous germline *SDHB* mutation from which RS0 and RS1/2 were isolated, there is a new *SDH*-deficient PGL model derived from mice. Multiple PHEOs arise in mice in which complete loss of *SDHB* was combined with loss of *NF1* [31]. The *SDHB/NF1* mouse model has provided insights into early mechanisms of tumorigenesis, but derivative cell lines have not yet been established. Other new investigative tool includes the MENX rat that carries a frameshift mutation in the *CDKN1B* gene (encoding for p27) and spontaneously develops PHEO. Intriguingly, these tumors recapitulate most characteristics of *SDH*-related PGL including norepinephrine and dopamine secretion, a *HIF-2α*-driven pseudohypoxic signature, metabolomics reprogramming associated with an accumulation of the oncometabolite 2-hydroxyglutarate, DNA hypermethylation and massive angiogenesis [32]. A zebrafish model in which loss of *SDHB* in homozygous larvae recapitulates the metabolic characteristics of human PGL has also been generated, but tumors have not so far been observed in these animals [33].

20.4 Complex 3D Adrenal *In Vitro* Preclinical Models

Given the physiological relevance of the functional interaction occurring between the steroidogenic cortical and the chromaffin medullary components of the adrenal gland, it is likely that the crosstalk that regulates the gland organogenesis and functions [34] may also affect the pathogenesis and progression of adrenal tumors. The peculiar organization of mammalian adrenals with a portal system connecting cortex and medulla strongly supports the evolutionary importance of the crosstalk inside the same gland. Local glucocorticoids have been demonstrated to stimulate medulla activity, in particular by upregulating the key enzymes in catecholamine biosynthesis [34], while medullary neuroendocrine peptides support cortical secretion and growth [35]. Therefore, a relevant implementation of the 3D *in vitro* models for these two types of cancers should consider not only the TME but also the other secretory part:

- *Ex vivo fetal adrenal gland explants or adrenal organoids derived in vitro from mixed primary cell populations obtained from fetal gland cell dissociation and reassembly.* The complex gland organization and secretory activity are maintained in the explant model [36] but also in the *in vitro*-induced organoids derived from primary mixed cell populations isolated from fetal adrenals, which display a spatial organization and a steroidogenic and catecholamine secretion resembling the gland of origin [37].
- *Tumor tissue slice cultures* have recently been developed as innovative tools for *in vitro* testing drug efficacy for personalized treatment strategy [38]. Tumor specimens left intact by mechanically slicing after surgical removal are incubated under standardized organo-culture conditions in either static or microfluidic settings up to several days to test different drug treatments.
- *Mixed spheroids* obtained from *in vitro* co-culture of NCI-H295R and MTT cell lines have been recently developed. These 3D structures, called adrenoids, mimic the organization of the gland of origin and the distinct endocrine activity of medullary catecholamine and cortex corticosteroid secretion, demonstrating a growth advantage due to the coexistence of the two endocrine components [39].

References

1. Rosfjord E, Lucas J, Li G, Gerber HP. Advances in patient-derived tumor xenografts: from target identification to predicting clinical response rates in oncology. Biochem Pharmacol. 2014;91(2):135–43.
2. Sedlack AJH, Hatfield SJ, Kumar S, et al. Preclinical models of adrenocortical cancer. Cancers (Basel). 2023;15(11):2873.
3. Sigala S, Rossini E, Abate A, et al. An update on adrenocortical cell lines of human origin. Endocrine. 2022;77(3):432–7.

4. Kurlbaum M, Sbiera S, Kendl S, et al. Steroidogenesis in the NCI-H295 cell line model is strongly affected by culture conditions and substrain. Exp Clin Endocrinol Diabetes. 2020;128(10):672–80. Erratum in: Exp Clin Endocrinol Diabetes. 2020;128(10):e3.
5. Ragazzon B, Lefrançois-Martinez AM, Val P, et al. Adrenocorticotropin-dependent changes in SF-1/DAX-1 ratio influence steroidogenic genes expression in a novel model of glucocorticoid-producing adrenocortical cell lines derived from targeted tumorigenesis. Endocrinology. 2006;147(4):1805–18.
6. Ragazzon B, Lefrançois-Martinez AM, Val P, et al. ACTH and PRL sensitivity of highly differentiated cell lines obtained by adrenocortical targeted oncogenesis. Endocr Res. 2004;30(4):945–50.
7. Armignacco R, Cantini G, Poli G, et al. The adipose stem cell as a novel metabolic actor in adrenocortical carcinoma progression: evidence from an in vitro tumor microenvironment crosstalk model. Cancers (Basel). 2019;11(12):1931.
8. Cerquetti L, Bucci B, Raffa S, et al. Effects of sorafenib, a tyrosin kinase inhibitor, on adrenocortical cancer. Front Endocrinol (Lausanne). 2021;12:667798.
9. Nilubol N, Zhang L, Shen M, et al. Four clinically utilized drugs were identified and validated for treatment of adrenocortical cancer using quantitative high-throughput screening. J Transl Med. 2012;10:198.
10. Bornstein S, Shapiro I, Malyukov M, et al. Innovative multidimensional models in a high-throughput-format for different cell types of endocrine origin. Cell Death Dis. 2022;13(7):648.
11. Hantel C, Shapiro I, Poli G, et al. Targeting heterogeneity of adrenocortical carcinoma: Evaluation and extension of preclinical tumor models to improve clinical translation. Oncotarget. 2016;7(48):79292–304.
12. Kiseljak-Vassiliades K, Zhang Y, Bagby SM, et al. Development of new preclinical models to advance adrenocortical carcinoma research. Endocr Relat Cancer. 2018;25(4):437–51.
13. Abate A, Tamburello M, Rossini E, et al. Trabectedin impairs invasiveness and metastasis in adrenocortical carcinoma preclinical models. Endocr Relat Cancer. 2022;30(2):e220273.
14. Ruggiero C, Tamburello M, Rossini E, et al. FSCN1 as a new druggable target in adrenocortical carcinoma. Int J Cancer. 2023;153(1):210–23.
15. Tamburello M, Abate A, Rossini E, et al. Preclinical evidence of progesterone as a new pharmacological strategy in human adrenocortical carcinoma cell lines. Int J Mol Sci. 2023;24(7):6829.
16. Basham KJ, Hung HA, Lerario AM, Hammer GD. Mouse models of adrenocortical tumors. Mol Cell Endocrinol. 2016;421:82–97.
17. Berthon A, Sahut-Barnola I, Lambert-Langlais S, et al. Constitutive beta-catenin activation induces adrenal hyperplasia and promotes adrenal cancer development. Hum Mol Genet. 2010;19(8):1561–76.
18. Wilmouth JJ Jr, Olabe J, Garcia-Garcia D, et al. Sexually dimorphic activation of innate antitumor immunity prevents adrenocortical carcinoma development. Sci Adv. 2022;8(41):eadd0422.
19. Bayley JP, Devilee P. Advances in paraganglioma-pheochromocytoma cell lines and xenografts. Endocr Relat Cancer. 2020;27(12):R433–50.
20. Martinelli S, Maggi M, Rapizzi E. Pheochromocytoma/paraganglioma preclinical models: which to use and why? Endocr Connect. 2020;9(12):R251–60.
21. Tischler AS, Favier J. Progress and challenges in experimental models for pheochromocytoma and paraganglioma. Endocr Relat Cancer. 2023;30(5):e220405.
22. Greene LA, Tischler AS. Establishment of a noradrenergic clonal line of rat adrenal pheochromocytoma cells which respond to nerve growth factor. Proc Natl Acad Sci USA. 1976;73(7):2424–8.
23. Hopewell R, Ziff EB. The nerve growth factor-responsive PC12 cell line does not express the Myc dimerization partner Max. Mol Cell Biol. 1995;15(7):3470–8.
24. Martiniova L, Lai EW, Elkahloun AG, et al. Characterization of an animal model of aggressive metastatic pheochromocytoma linked to a specific gene signature. Clin Exp Metastasis. 2009;26(3):239–50.
25. Powers JF, Evinger MJ, Tsokas P, et al. Pheochromocytoma cell lines from heterozygous neurofibromatosis knockout mice. Cell Tissue Res. 2000;302(3):309–20.

26. D'Antongiovanni V, Martinelli S, Richter S, et al. The microenvironment induces collective migration in SDHB-silenced mouse pheochromocytoma spheroids. Endocr Relat Cancer. 2017;24(10):555–64.
27. Letouzé E, Martinelli C, Loriot C, et al. SDH mutations establish a hypermethylator phenotype in paraganglioma. Cancer Cell. 2013;23(6):739–52.
28. Powers JF, Cochran B, Baleja JD, et al. A xenograft and cell line model of SDH-deficient pheochromocytoma derived from Sdhb+/− rats. Endocr Relat Cancer. 2020;27(6):337–54. Erratum in: Endocr Relat Cancer. 2020;27(10):X9–X10.
29. Ghayee HK, Bhagwandin VJ, Stastny V, et al. Progenitor cell line (hPheo1) derived from a human pheochromocytoma tumor. PLoS One. 2013;8(6):e65624.
30. Matlac DM, Hadrava Vanova K, Bechmann N, et al. Succinate mediates tumorigenic effects via succinate receptor 1: potential for new targeted treatment strategies in succinate dehydrogenase deficient paragangliomas. Front Endocrinol (Lausanne). 2021;12:589451.
31. Armstrong N, Storey CM, Noll SE, et al. SDHB knockout and succinate accumulation are insufficient for tumorigenesis but dual SDHB/NF1 loss yields SDHx-like pheochromocytomas. Cell Rep. 2022;38(9):110453.
32. Mohr H, Ballke S, Bechmann N, et al. Mutation of the cell cycle regulator p27kip1 drives pseudohypoxic pheochromocytoma development. Cancers (Basel). 2021;13(1):126.
33. Dona M, Waaijers S, Richter S, et al. Loss of sdhb in zebrafish larvae recapitulates human paraganglioma characteristics. Endocr Relat Cancer. 2021;28(1):65–77.
34. Bechmann N, Berger I, Bornstein SR, Steenblock C. Adrenal medulla development and medullary-cortical interactions. Mol Cell Endocrinol. 2021;528:111258.
35. Haase M, Willenberg HS, Bornstein SR. Update on the corticomedullary interaction in the adrenal gland. Endocr Dev. 2011;20:28–37.
36. Melau C, Nielsen JE, Perlman S, et al. Establishment of a novel human fetal adrenal culture model that supports de novo and manipulated steroidogenesis. J Clin Endocrinol Metab. 2021;106(3):843–57.
37. Poli G, Sarchielli E, Guasti D, et al. Human fetal adrenal cells retain age-related stem- and endocrine-differentiation potential in culture. FASEB J. 2019;33(2):2263–77.
38. Seliger B, Al-Samadi A, Yang B, et al. In vitro models as tools for screening treatment options of head and neck cancer. Front Med (Lausanne). 2022;9:971726.
39. Martinelli S, Cantini G, Propato AP, et al. The 3D in vitro Adrenoid cell model recapitulates the complexity of the adrenal gland. Sci Rep. 2024;14(1):8044.